AN INTRODUCTION TO ABSTRACT ALGEBRA

AN INTRODUCTION TO ABSTRACT ALGEBRA

VOLUME 1

BY

F. M. HALL

Head of the Mathematics Faculty,
Shrewsbury School

SECOND EDITION

CAMBRIDGE UNIVERSITY PRESS

Cambridge

London New York New Rochelle

Melbourne Sydney

CAMBRIDGE UNIVERSITY PRESS
Cambridge, New York, Melbourne, Madrid, Cape Town, Singapore, São Paulo, Delhi

Cambridge University Press
The Edinburgh Building, Cambridge CB2 8RU, UK

Published in the United States of America by Cambridge University Press, New York

www.cambridge.org
Information on this title: www.cambridge.org/9780521084840

First published 1966
Second edition 1972
First paperback edition 1980
Reprinted 1980
Re-issued in this digitally printed version 2008

A catalogue record for this publication is available from the British Library

Library of Congress Catalogue Card Number: 75-185565

ISBN 978-0-521-08484-0 hardback
ISBN 978-0-521-29861-2 paperback

CONTENTS

PREFACE

This work, to be completed in volume 2, is written at a time when abstract algebra is being introduced increasingly into the schools. It attempts to give a broad introduction to the subject, and is intended for those with no previous knowledge of this work but with a fair amount of mathematical sophistication. The level at which the book is written is that of a fairly intelligent sixth-former who wishes to know something about modern algebra and the work that leads up to it, and it could well be read by such a boy before he enters a university. Volume 2 in particular should be useful also for first-year university students, as a general background before a detailed study of the various branches of algebra, while teachers of mathematics who have not studied abstract algebra themselves but who, nevertheless, wish to learn about it and possibly to teach it should find parts of both volumes interesting and useful. While books on groups, rings, vector spaces and the other topics abound, it is not easy to find many which start from the beginning and lead up to the ideas gradually and in a fairly elementary manner. It is hoped that in this the book will satisfy a need.

The present volume leads up to the abstract ideas and methods by means of the study of various particular cases. After a little work on general sets and set theory it deals with the special sets of the integers, other number sets, residues, polynomials and vectors. These should be fairly familiar to the reader, but the emphasis is on those properties that carry over into more general abstract structures, and the proofs selected bear this in mind. Some may wish to omit parts of these chapters as being already known, but some of the results, though important and fairly simple, are not easily available in ordinary text-books.

After a chapter on mappings we study in detail the fundamental laws of algebra, which have lain behind the previous work, and then the final three chapters introduce the theory of groups, give plenty of examples, and study the idea of subgroups as far as Lagrange's theorem.

Volume 2 will continue group theory with a chapter on group homomorphisms and will then introduce elementary ideas in the study of rings, fields and integral domains. Invariant subgroups and ideals will be discussed, and there will be a chapter on vector spaces in which matrices will be mentioned, though no detailed account of matrix theory or linear algebra will be given, since the methods and results of these subjects are, I believe, different in kind to those of abstract algebra proper, being analytic rather than synthetic, more concerned with the properties of individual elements than with the structure as a whole. Volume 2 will end with more detailed work on the algebra of sets and Boolean algebra, and an indication of the main ways in which the work of the whole book is developed into more advanced topics.

Throughout the book I have been careful to give detailed explanations of the reasons for the work, and of the methods used. The technical language has been kept within bounds, as has the symbolism. Yet the work is rigorous as far as it goes, and the notation is in accordance with normal usage, though there is no general agreement in this respect. I have explained new notation as it arises, and occasionally have used my own, as in chapter 6 where residues are printed in bold type. The reader will have nothing to 'unlearn' when he passes on to more advanced text-books.

At each stage I have given as many concrete examples of the structures as I could. It is not always easy to find convincing ones (for example, most simple illustrations of Venn diagrams could be understood just as easily without their aid), and many are taken from other branches of mathematics, but I have done my best and the stock of examples will increase as the subject is taught more and more at an elementary level.

The book is intended to be read with little or no aid from a teacher (not that such aid should be scorned if available) and each chapter ends with a few worked exercises. The exercises themselves are divided into A and B: the first are quite straightforward and should be worked completely. The B exercises are very variable; some are fairly straightforward, others quite difficult and a few give extensions of the bookwork. The reader is not expected to be able to do all these exercises, at least not at a first reading. As with practical examples, so with exercises it is not easy to find those which, without being impossible for any but research workers, are yet non-trivial and worthy of the attention of the student. Here again I have done my best, and here also the stock should increase with use.

I would like to thank some of my former pupils at Dulwich College who read the manuscript and made valuable suggestions. I am indebted to my colleague, Mr D. B. Pennycuick, who read the proofs; and am grateful to the Cambridge University Press for their help throughout all stages of the preparation of the book.

Dulwich College F. M. H.
December 1964

1

INTRODUCTION

1.1. The nature of algebra

Algebra is first a generalisation of arithmetic. Instead of dealing with particular numbers we use letters to denote arbitrary ones and work with these according to the usual rules of arithmetic. It is concerned with those properties and processes that are common to all numbers and not those, such as primeness, peculiar to certain numbers or integers. The interest of algebra lies in the processes we use and their consequences, some of the chief fields of the work being the use of the four rules of addition, subtraction, multiplication and division and problems involving them, such as simplification of expressions, the solution and investigation of equations, the study of polynomials and other functions and their graphs, and the investigation of inequalities.

Elementary algebra uses letters to stand for numbers of various types: fractions, real numbers and later complex numbers. Most of the work is similar in all these cases, the processes and rules being almost identical, and the algebra is not basically concerned with the particular set of numbers in question, but rather with the methods and rules for combining them.

Algebra is essentially a finite process. We often include under the heading of 'algebra' such topics as convergence of series and the study of transcendental functions such as the exponential and logarithmic functions, but these properly belong to analysis, which is concerned with limiting processes and the infinite and infinitesimal.

1.2. Abstract algebra

As we have indicated, algebra is concerned basically with the processes and rules of combination of numbers, rather than with the numbers themselves. This was not realised by the early workers, but in the early part of the nineteenth century mathematicians gradually came to understand that the same or similar processes to those used in elementary algebra could be applied to many objects or sets other than numbers, and modern abstract algebra came into being.

Modern algebra then is concerned with sets of objects and possible ways of combining the elements of the set. Many rules of combination have been investigated and others are still being studied: the most fruitful are those with similar properties to the ordinary four rules of elementary work, though applied much more generally than to numbers alone.

Sets, together with rules of combining their elements, form *algebraic structures*. Much of the interest is synthetic, i.e. is concerned with the shape of the structure as a whole, but the analytical aspect of investigation of the elements themselves is important in some cases.

In this book we deal first with various special structures, emphasising those aspects that are of general application, and later, particularly in volume 2, investigate more general structures. Thus the reader is led gradually into the purely abstract work, and even there many concrete examples are given.

1.3. The axiomatic approach to mathematics

Modern pure mathematics is almost entirely axiomatic in approach. This is a fairly recent development: for most of its history mathematics has been ostensibly based on the natural or everyday world. Thus, whatever Euclid himself believed his system of geometry to be, it has usually been taken to be a description of physical space, while the nature of real numbers was held to be self-evident by most people. Such foundations gradually proved unsatisfactory. The assumptions behind Euclid are difficult to state clearly and since the formulation of the theory of relativity it has been found that the natural world does not obey them anyway. Analysis became rigorous only about 200 years ago, and the nature of irrational numbers was not described until 1872, while the nature of number itself is still being discussed.

The intuitive ideas behind mathematics are thus not secure, and the study of the foundations belongs more properly to philosophy: mathematics itself is concerned with the deductions obtained from the basic ideas. It is therefore more satisfactory to lay down certain initial axioms or postulates and to deduce from them according to the accepted laws of logic (which are themselves the subject of study by philosophers). Mathematics is not concerned with the question whether the axioms are 'true' or not; all it can say is that given certain assumptions then other results and consequences follow logically. Theoretically any set of axioms may be chosen provided it is self-consistent, but obviously the work will be unfruitful unless the choice is a careful one, and natural phenomena and the traditional fields of study can lead us to suitable sets of axioms. Thus Euclid may be put on a proper footing by assuming certain axioms, while the choice of similar sets will lead to the various non-Euclidean geometries. The fewer the axioms the greater generality the resulting system possesses, but the fewer the results that may be deduced.

Thus modern mathematics lays down postulates and deduces from them. This can be a surprisingly fruitful process, both in practical terms (since many practical systems will obey the axioms chosen) and in aesthetic ones. It enables us to study systems that seem at first sight impossible but which often turn out to be extremely useful. For example, the study of space of four dimensions would seem useless at first sight, but it is vital to relativity theory and also in electromagnetism.

Abstract algebra lays down postulates for combining elements of sets and studies their consequences. For numbers these are the fundamental laws of addition and multiplication (the Commutative, Associative and Distributive Laws), while the choice of some only of these leads to the study of more general structures. A surprising amount of work may be done with very few axioms in this subject (group theory has only three basic laws but research is still very active in the subject).

1.4. Logic in mathematics

Mathematics uses the laws of logic and we do not attempt to lay down what these are or to study them. There are, however, a few important logical ideas which are often not understood properly by the mathematical student but which are vital to much of his work, especially that which is concerned with the consequences of axioms rather than the techniques of manipulation. We explain some of these here.

Equality and identity

In elementary work the equals sign is usually used to indicate that two expressions have the same value, e.g. $x+2 = 4$ means that $x+2$ and 4 have the same value, for some particular value of x, which often has to be found. If two expressions have the same value for all values of the variable concerned we usually use the identity sign; thus $(x+2)^2 \equiv x^2+4x+4$. The distinction is often blurred in practice. The equals sign is also used as a special case of an inequality, thus we use '$\leqslant$' and '$\geqslant$', and both being true implies equality.

We will use '$a = b$' to mean that a and b *are the same*. This implies nothing about inequalities and may be used whatever type of element we are dealing with. Thus for numbers '$x = y$' means that x and y are the same number, while if we are dealing with polynomials '$P(x) = Q(x)$' means that $P(x)$ and $Q(x)$ are the same polynomial, according to our definition of sameness of polynomials (they are of the same degree and have all coefficients the same). If A and B are sets '$A = B$' means that they are the same set, not merely that they are of equal size. Other uses of the symbol (for example, for isomorphic groups) will be given as we require them.

Implication

If a statement A leads logically to another statement B we say that 'A implies B' and write $A \Rightarrow B$. For example: R is a square $\Rightarrow$ R is a rectangle, or $x = y \Rightarrow x^2 = y^2$, or the positive integer n is even $\Rightarrow$ n can be divided by 2. If $A \Rightarrow B$ then B is implied by A and we sometimes write $B \Leftarrow A$. If it does not follow logically that B is true when A is, we write $A \not\Rightarrow B$ or $B \not\Leftarrow A$.

In the first two examples above we have $A \Rightarrow B$ but $B \not\Rightarrow A$, while in the third $B \Rightarrow A$ also. In this case we say that 'A implies and is implied by B' and write $A \Leftrightarrow B$. In such a case the statements are logically equivalent and the argument in which B follows A is reversible. This is not the case in the other two examples. Again the distinction is often not realised, as when solving an equation we square both sides. This is not reversible and so, although any solution must be given by our method, it does not follow that a solution we obtain is a solution of the original equation, and each must be checked.

If A and B must be either true or false statements, so that if we call the negation of A 'not A' then either A or not A is true, $A \Rightarrow B$ is logically equivalent to 'not $B \Rightarrow$ not A'.

The sign $\Rightarrow$ must not be confused with $\rightarrow$, nor $\Leftrightarrow$ with $\leftrightarrow$, the latter being used for various correspondences. Thus in inversion we could write $P \rightarrow P'$ or $P \leftrightarrow P'$ where P and P' are inverse points.

Necessary and sufficient conditions

A is a *necessary* condition for B means that $B \Rightarrow A$: if B is true then A must be.

A is a *sufficient* condition for B means that $A \Rightarrow B$: if A is true then B must be.

Thus a necessary condition for an integer greater than 2 to be prime is that it is odd, but this is not sufficient. A sufficient condition for a figure to be a rectangle is that it is a square, but this is not necessary. However, the necessary and sufficient condition for two triangles to have their sides in proportion is that they are equiangular.

We often have a set of necessary and sufficient conditions. (A is a *necessary and sufficient* condition for B means that $A \Leftrightarrow B$.) Thus necessary and sufficient conditions for a figure to be a square are that it is a rectangle and also a rhombus.

If and only if

This gives another way of thinking about the ideas of implication. If $A \Rightarrow B$ we say that B is true *if* A is true, while if $B \Rightarrow A$ we say that B is true *only if* A is. Thus 'B true if A is' means that A is a sufficient condition for B and vice versa, while 'B true only if A is' means that A is a necessary condition for B. If $A \Leftrightarrow B$ we say that B is true *if and only if* A is true.

Counter examples

If we have a theorem which seems likely to be true but which may not be, usually the best way of proving it false is to find an example where it is not true. Such an example is called a counter example, the German 'Gegenbeispiel' being sometimes used. A famous instance is Fermat's theorem on binary powers, which states that the number $2^{2^n}+1$ is prime for all n. Although Fermat believed in the truth of this theorem he could not prove it, and in 1732 Euler discovered that if $n = 5$ the number is composite: this counter example of course immediately disproves the theorem.

As a further example, it is easily proved that if Σu_r is convergent then $u_r \to 0$, and we may think that the converse is true, but $u_r = 1/r$ gives a counter example.

If after due consideration we cannot find a counter example we may reasonably suppose that the theorem is true, but this of course is not proved, and it may well be false in some obscure cases. Goldbach's conjecture, that every even integer may be expressed as the sum of two primes, has never been proved, but no counter example has been discovered and most mathematicians believe in the truth of the conjecture.

Reductio ad absurdum

A common way of proving a theorem is to assume that it is false and then to show that this leads to a logical contradiction or to an obviously false result. For example, to prove that there is no greatest prime we assume that there is and let the greatest prime be n. Then $n!+1$ (where $n!$ means the product $n(n-1)(n-2)\ldots 2.1$) either is a prime greater than n or has a prime factor greater than n (all integers from 2 to n are factors of $n!$ and so cannot be factors of $n!+1$), and in either case we have a contradiction of our supposition. This method is particularly common in proving a converse: the theorem that if the opposite angles of a quadrilateral are supplementary then the quadrilateral is cyclic is usually proved in this way, assuming the basic theorem that the opposite angles of a cyclic quadrilateral are supplementary.

1.5. Historical summary

The first notable algebraists were the Arabs. The Egyptians, Greeks and Hindus had all done a little work in this subject, but the Arabs were the first to concentrate on it, mainly in connection with astronomy, and progressed so far as the solution of cubic equations. The word 'algebra' is a corruption of the Arabic 'al-jebr', meaning the transposing of negative terms in an equation to the other side.

At the time of the Renaissance, algebra became one of the main fields of mathematical study. Cubics were solved for the general case by Tartaglia (about 1499–1557) and Cardan (1501–76) and quartics by Ferrari (1522–65). Vieta (1540–1603) introduced letters to stand for unknown quantities, while the symbols + and − appear first in a book printed in 1489, and the exponential notation for powers was introduced by Descartes (1596–1650).

Newton (1642–1727) worked on the theory of equations and the binomial theorem, and about this time negatives came to be accepted as proper numbers. Complex numbers were also used but were imperfectly understood until later. Argand's famous paper on the geometrical interpretation of imaginary quantities was published in 1806, while Gauss finally put complex numbers on an equal footing with the real numbers in 1831. Gauss gave the first fully satisfactory proof of the 'Fundamental Theorem of Algebra' that a polynomial equation of the nth degree has exactly n roots, his first proof being discovered in 1797.

Determinants were studied by Wronski (1778–1853), Cauchy (1789–1857) and Jacobi (1804–51) among others, while matrices were introduced at about the same period, much of the work being by Hamilton (1805–65) and Cayley (1821–95). Hamilton also invented quaternions, the first non-commutative system to be studied intensively, which were superseded for practical purposes by matrices and tensors. The theory of invariants and linear transformations, connected with matrix theory and leading to modern linear algebra, was developed by Cayley and Sylvester (1814–97) and by Hermite (1822–1901).

A milestone in the development of modern abstract ideas was

Boole's publication of *The Laws of Thought* in 1854. This applied mathematics to logic and marked the first real break from traditional ideas, based on practical ideas of number and space. Boole's work showed that algebra is not necessarily concerned with numbers but that the processes may be used much more generally.

The beginnings of the ideas of group theory lay in the solubility of equations. Quartics had been solved by Ferrari in the sixteenth century but nobody had been able to give a solution for the general quintic, and this was finally proved impossible by Abel in 1826. Galois simplified his solution in about 1830 and discovered a great deal about groups in connection with the solution of equations, besides investigating invariant subgroups and the theory of fields. He was the first to use the word 'group' in the modern sense. The theory was elaborated by Lagrange, Cayley and particularly Cauchy (in about 1844–46). At this early period groups were thought of in terms of permutations or substitutions, or sometimes in connection with residues (Euler) and number theory. Definitions of abstract groups were given by Kronecker in 1870 and later simplified. Other notable early workers in group theory were Jordan (composition series and conditions for groups to be soluble), Sylow (1832–1918) (subgroups), Sophus Lie (1842–99) (topological groups) and Klein (groups of the regular polyhedra).

Topology, originally known as 'analysis situs', was studied by Euler and others, but was only gradually recognised as a separate subject, distinct from geometry.

In the present century the growth of abstract ideas has been rapid. Many algebraic systems have been studied and research is still active both in the 'pure' algebra of groups, rings and fields and more recently in algebra applied to topological structures.

2

SETS

2.1. The idea of a set

A set is merely a collection of objects. Although the idea is basically simple, and indeed hardly seems to need stating, it is the most important concept in mathematics. (The latter has even been described as being the study of various aspects of set theory!) The reason is that mathematics is essentially a process of abstraction—we select certain properties of the objects with which we are working and apply the laws of logic to deduce further properties. We cannot do this without putting some restriction on our objects: we deal, in a certain piece of work, only with objects which are in a given set.

It may not at first sight appear obvious that elementary mathematics restricts itself in this way. We tend to think of arithmetic as applying to everything, but of course this is a false idea. Arithmetic in the first instance is concerned merely with the properties of numbers—we do it in the *set* of numbers. Even here we work in different sets at different stages. At first we restrict ourselves to the set of positive whole numbers, which we later extend to include fractions, then negatives, and finally we do our arithmetic within the set of all real numbers. When we apply our arithmetic to problems we extend the sets that we use to include, for instance, all objects which have a monetary value, or all baths with two taps (in the famous calculations of this type). Notice, however, that the properties which we abstract from these sets are precisely those which are possessed by ordinary numbers. There is no new mathematics involved, and so these sets of practical objects have little purely mathematical interest.

When we start algebra we still, in the elementary stages, keep within the set of numbers. We let x 'stand for' any number. We very easily lose sight of the basic set, and later we start using our letters to stand for any *complex* number, thus extending our real number domain of operations to the complex domain. Remarkably our algebra is nearly the same, and the purpose of this book is to show how we can still keep the same algebra, or at any rate some of it, when working in sets other than the real or complex numbers.

The study of geometry leads us into new sets. We work, in Euclid, within any one of the set of planes, we deal with the set of points in that plane, and with the sets of lines and triangles. Our results apply to the sets of objects which satisfy certain postulates.

In advanced mathematics more varied sets are encountered. Differentiation can be performed only within the set of *differentiable* functions (i.e. functions which possess a derivative at a certain point, a property which by no means all functions have). There is a separate set of *integrable* functions. We are interested in the set of convergent series—we cannot speak of the sum of a series that isn't convergent. The method of induction applies only to the set of integers. We make excursions into the set of vectors.

The above examples are of mathematical sets, but of course the idea may be applied to any collection of objects. The objects are called *elements*. They may be of any type, and even of varied types. The set may consist of a finite number of elements, or of infinitely many. We may not even know how many. So long as we can say of any object that it is either an element or is not, then we have defined a set. Thus we may consider the set of all mammals who have been parents of live-born young that have ever lived. It would be difficult to give an estimate of the size of this set but, given any object, it is possible to tell whether or not it is in the set. (At least it is possible in theory, provided we have an exact definition of mammal.) No object that is not a mammal need be considered and every mammal either has been a parent or has not. Notice that the definition is precise.

The above example has a simple definition, but this need not be so. Any selection of, say, ten thousand insects forms a set, and there may be no obvious connection between the elements in this case. We may even take ten thousand insects and one jam-jar and thus form another set. The elements may even be sets themselves: for example, we could have the set of all sets of insects. The important point is that any object we name must be either in the set or not in it: in other words, the set must be precisely defined.

It is seen that the idea of a set is exceedingly wide, and it seems strange that it is a fruitful concept. We can do a surprising amount of theory with sets in general (some will be done in this chapter and the next), but of course most mathematics deals with, or within, some specified set or type. Some particularly important sets will be studied in chapters 4–8, while the remainder of this volume, and volume 2, deal with some types which are important from the algebraic point of view.

2.2. Notation for sets

In any branch of mathematics we find that certain operations occur over and over again, and we introduce special notations for these in order to simplify our work. Thus in arithmetic we have the four symbols $+$, $-$, $\times$, $\div$ to stand for the four basic processes. Such symbolism is a necessity for proper mathematical progress; without its use our work is complicated so much by unnecessary words that it cannot develop to any extent, while on the other hand a good notation in any branch of the subject is a tremendous help both for the mathematics and for our understanding and appreciation of it.

In set theory as in the other branches of mathematics there is a certain amount of notation which we use and which it is necessary to learn and become competent in handling. Since sets enter into so much of all mathematics this notation is met with throughout the subject in its more advanced parts, and for this reason it is of particular importance that it be known thoroughly. We will meet with it throughout this book.

In this section we lay down explicitly the basic notations used. Although many of the symbols are used more or less universally there is no general agreement between different authors, and there is often some variation in the meanings to be attached to certain parts of the notation. We will note any important divergences of usage, but it is always wise when reading a book by a strange author to check his meanings.

Sets and elements. Equality

In general sets are denoted by capitals and elements by small letters, although there are many exceptions to the latter statement owing to the very wide variety of possible types of elements.

If A and B are two sets, $A = B$ means that A and B consist of precisely the same elements, in other words are the same set. They do not merely have the same number or type of elements, or have the same properties.

$A \neq B$ means that A and B are not the same set.

Membership

The most important thing to know about a set is what are its members, i.e. is a given object a an element of the set A or not.

$a \in A$ means that a is an element of the set A and is read as 'a is a member of A' or 'a is in A'.

$A \ni a$ means the same as the above and is read 'A contains a'.

The difference between the above two notations is that in the first the emphasis is on the element a, while in the second the emphasis is on the set A.

$a \notin A$ and $A \not\ni a$ both mean that a is not in A.

Note that an oblique or vertical line through a symbol is used generally to mean its negation.

Inclusion

If A and B are two sets, $A \subseteq B$ means that A is contained in B, i.e. every member of A is a member of B but not, of course, *necessarily* vice versa, although this may be true also.

$A \subset B$ means that $A \subseteq B$ but $A \neq B$. This means that although every element of A is also an element of B, there exist some elements of B which are not in A: B is 'bigger' than A.

Note that $A \subseteq B$ includes the two cases $A \subset B$ and $A = B$.

Important Note. The above is the meaning we will attach to the symbols $\subseteq$ and $\subset$, but many authors use $\subset$ to include the possibility of equality. Thus they use $A \subset B$ to mean 'A is contained in B' and do not use the symbol $\subseteq$. Our usage seems more natural and helps to bring out the analogy with $\leqslant$ and $<$ as used in the algebra of real numbers.

$A \supseteq B$ means that A contains B, i.e. that every element of B is an element of A, while $A \supset B$ means that A strictly contains B, i.e. that $A \supseteq B$ and $A \neq B$. Note the distinction between $A \supseteq B$ (where B is a set) and $A \ni a$ above, where a is an element.

Note that $A \supseteq B$ means that $B \subseteq A$, in the same way that $x \geqslant y$ and $y \leqslant x$ mean the same when x and y are real numbers.

$A \nsubseteq B$ means that A is not contained in B.

Notice that $A \subseteq B$ and $A \supseteq B$ together imply that $A = B$: this is a common way of proving two sets equal.

Union and intersection

In the same way that we combine two numbers together by one of the four rules to form a new number (i.e. $3 \times 5 =$ the new number 15), we can combine two or more sets in various ways to form a new set. The processes themselves form the subject of chapter 3, but we will give the basic notation here.

$A \cup B$ means the set of elements which are in *either* (or both) of the sets A and B, and is read 'A union B' or, colloquially, 'A cup B'.

$A \cap B$ means the set of elements, which are in *both* of the sets A and B and is read 'A intersection B' or 'A cap B'.

We extend the above to three or more sets. Thus $A \cup B \cup C$ is the set of elements in one or more of the sets A, B or C, while $A \cap B \cap C$ is the set of elements which are in all three sets. If we wish to speak of the union of the n sets $A_1, A_2, \ldots, A_n$ we may write either $A_1 \cup A_2 \cup A_3 \cup \ldots \cup A_n$ or, better, $\bigcup_{i=1}^{n} A_i$, and similarly for intersection as $A_1 \cap A_2 \cap A_3 \cap \ldots \cap A_n$ or $\bigcap_{i=1}^{n} A_i$.

The way to remember which symbol stands for union and which for intersection is to think of $\cup$ as being u, the initial letter of union.

The empty set and the universal set

It is often convenient to speak of the set which contains no elements at all. It seems strange to think of this as a set in the sense of the latter being a collection of elements, but it fulfils the basic requirement that we know whether any given object is an element or not, for no object is an element of this set. The set is called the *empty set* or *null set* and is denoted by $\emptyset$.

One reason for the necessity of considering the empty set is as follows. Given any two sets A and B then $A \cap B$ is always a set, or at least we wish it to be so. If A and B have elements in common then their intersection is an obvious set, but if they are mutually exclusive then we must say that their intersection is the empty set, i.e. we say that $A \cap B = \emptyset$.

The set that contains all the possible objects which might enter into a particular problem or piece of work is called the *universal set* for that problem or work. For example, in real arithmetic the universal set would be the set of all real numbers, while in work on a population census it would be the set of all human beings in the country in question. It is usually obvious in any problem what the universal set is, but in cases of doubt it is stated explicitly. Various notations are used for it, but as we will rarely have to specify such a set we do not give any specific symbol here.

Difference and complement

There is no obvious meaning for $A+B$, but we use $A-B$ to mean the set of all elements in the universal set in question which are in A but not in B. It does not necessarily imply that $B \subseteq A$. Thus if A is the set of all male rabbits and B the set of all white rabbits, working in the universal set of all rabbits, then $A-B$ is the set of all male rabbits that are not white.

The set of all elements in the universal set that are not elements of A is called the *complement of* A and is written either as $\bar{A}$ or as A'. In the above example $\bar{A}$ is the set of all female rabbits.

$\forall$ *and* $\exists$

We often wish to speak of a property as possessed by all the elements of a certain set, and would say: 'For all elements a in the set A, a has this property'. Since we know that A is a set and may assume a to be an element, this can be shortened to 'For all $a \in A$, a ...'. To shorten the statement further we need a symbol to stand for 'for all', and we use '$\forall$', an upturned A, to mean this. Thus the statement becomes: '$\forall a \in A$, a ...', and this is read 'For all a in A, a ...'.

Alternatively, we may wish to express the fact that a certain property is possessed by some element in A or, in other words, that A contains at least one element with the property. We could say: 'There exists an element a in the set A such that a has the property that ...' or, shortening, 'There exists $a \in A$ such that a ...'. We now introduce the symbol '$\exists$' to stand for 'there exists' and we also denote 'such that' by a full stop ., and the statement becomes: '$\exists a \in A . a$ ...', which is read 'There exists a in A such that a ...'.

A set expressed in terms of its elements

The set whose elements are the objects denoted by $a, b, c, \ldots, k$ is often expressed by writing the elements inside brackets thus: $\{a, b, c, \ldots, k\}$. Similarly, the set whose elements are $x_1, x_2, \ldots, x_n$ is written as $\{x_1, x_2, \ldots, x_n\}$ or, more briefly,

$$\{x_i : i = 1, \ldots, n\}$$

or even $\{x_i\}$.

If we have a set defined by a property, e.g. the set of all triangles, we may write it symbolically as $\{x: x \text{ is a triangle}\}$. The usefulness of this notation is more apparent in complicated examples. Thus the set of all functions which are continuous for all rational values of the independent variable x could be denoted by $\{f: f(x) \to f(a) \text{ as } x \to a \ \forall \text{ rational } a\}$.

Examples of the notation

Example 1. Suppose A is the set of all female rabbits, B that of all male rabbits, C all rabbits that are sons, D rabbits that are fathers and E all those that are grandfathers. (The universal set is that of all rabbits.)

If a is a particular baby male rabbit we see that $a \in B$ but $a \notin A$, while $C \ni a$ but $D \not\ni a$.

We also see that $E \subseteq D$ but $E \neq D$, so that in fact $E \subset D$. $E \subset B$ and $D \subset B$. $C \supset D$ and $A \not\supset D$, while $B = C$.

$D \cup E = D$, $A \cup B =$ the set of all rabbits, while $A \cap B = \varnothing$.

Example 2. Suppose A is the set of all dolls, B is the set of all dolls with blue eyes, C that of all those with fair hair and blue eyes, D all those with fair hair, E all those called Muriel, F all those with brown eyes, assumed non-empty. (Here A is also the universal set.)

Then $$C \subseteq B \subseteq A \quad \text{but} \quad C \neq B;$$

$$A \supseteq E, \quad \text{but} \quad B \nsupseteq E \quad \text{if} \quad F \cap E \neq \varnothing;$$

$$B \cap D = C, \quad \text{while} \quad B \cup D \subseteq A;$$

$$C \cap E \subseteq D, \quad C \cap F = \varnothing.$$

Although we cannot say that $E \subseteq F$ we cannot say either that $E \nsubseteq F$: we do not know whether or not there is a doll called Muriel which hasn't brown eyes. If, however, one such were exhibited we could say that $E \nsubseteq F$.

2.3. Subsets

The fact that $A \subseteq B$, where A and B are two sets, is more important than it seems at first sight. Suppose we have been working with the elements of B and have discovered certain properties of them and suppose later that we come across the elements of A. Then because $A \subseteq B$ all our properties of the elements of B hold automatically for those of A, and we already therefore know a fair amount about A.

For example, when we start work in the advanced branch of arithmetic known as the theory of numbers we deal almost exclusively with the positive integers and their properties. But the set of positive integers is contained in the set of real numbers, and so we can apply all our rules of manipulating the real numbers to them, in other words we can use all the ordinary methods of arithmetic. Of course the important results in the theory of numbers are those which apply only to the integers or to some of them, but in proving these results we find that we use many of the wider properties of the real numbers. What we have said is very obvious and there is no need to state it explicitly in our arithmetical work, but the justification, that the integers are contained in the reals, is important in spite of being self-evident.

The process described is often applied in the opposite direction. In dealing with the elements of B we often wish to restrict ourselves to those which have certain other properties. These will form a set A where $A \subseteq B$. Thus in arithmetic we may wish to restrict ourselves to the integers, and perhaps later we specialise still further with the even integer. In general the more we specialise the more properties we may discover, but of course the properties will not be of such wide application.

If $A \subseteq B$ we say that A is a *subset* of B: it consists of a part of B, possibly the whole. B is of course a subset of itself, while ø is a subset of all sets: these two are called the *trivial* subsets of B. All others are *non-trivial* subsets. All subsets other than B itself are sometimes called *proper* subsets.

We may have strings of sets, each a subset of the previous one. Thus let A_i consist of all integers which are multiples of 2^i. Then A_1 is all multiples of 2, A_2 is all multiples of 4, and so on, and we see that $A_1 \supseteq A_2 \supseteq A_3 \supseteq \ldots$. In this case the string is infinite, but we can of course have finite strings, as for example by terminating the above after any term.

2.4. Equivalence relations

We often have the elements of a set divided into several mutually exclusive subsets, every element belonging to one and only one of the subsets. Two examples are given below, and it is advisable to bear these concrete examples in mind as this section is read.

Example 1. The set of all living human beings may be divided into the subsets A_r, where A_r consists of all humans who are at the present time of age r complete years.

Example 2. The set of all integers, positive, negative or zero, may be divided into the subsets B_r, $r = 0, 1, \ldots, (n-1)$, where B_r consists of all those integers that leave remainder r when divided by the given positive integer n.

In any case such as the above let us consider the relationship between two elements which are in the same subset. We will denote the fact that y is in the same subset as x by the notation yRx, meaning that y stands in the given relationship (that of being in the same subset) to x. Then the relation R has the following properties.

(1) It is *reflexive*. xRx.

This means that for all x in our given set, x stands in the given relationship to itself, i.e. x is always in the same subset as itself, which is of course self-evident.

(2) It is *symmetric.* $yRx \Rightarrow xRy$.

If y stands in the relationship to x, then x does so to y: that is if y is in the same subset as x, then x is in the same subset as y.

(3) It is *transitive.* xRy *and* $yRz \Rightarrow xRz$.

If x is in the same subset as y, and y in the same one as z, then x and z are in the same subset.

The three properties above are necessarily true for the relation under consideration. The remarkable thing is that they are also sufficient. If we have a relation defined between certain pairs of elements of a set, which satisfies the three conditions above, then it divides the set into mutually exclusive subsets. Such a relation is called an *equivalence relation* and we now proceed to consider it more precisely.

Suppose we have a set S, and suppose there is a relation R between certain pairs of the elements of S. In order to define R all we need to know is whether any given pair of elements, taken in a certain order, stands in the relation R or not. We denote the fact that x and y, in that order, are in the relation R by the notation xRy.

R may be of any type, but the case that interests us is that which is reflexive, symmetric and transitive, in which case it divides the set S into mutually exclusive subsets. The proof of this important theorem is given below.

Theorem 2.4.1. *The equivalence classes theorem.*

Suppose in a set S we have a relation R defined between certain pairs of elements. Suppose further that R has the following 3 *properties:*

(1) *It is reflexive: i.e.* xRx *for all* $x \in S$.

(2) *It is symmetric: i.e.* $xRy \Rightarrow yRx$.

(3) *It is transitive: i.e.* xRy *and* $yRz \Rightarrow xRz$.

Then R is an equivalence relation: *i.e. it divides S into mutually exclusive subsets so that every element of S is in one and only one subset, and so that two elements are in the same subset if and only if they stand in the relation R to one another.*

The subsets are called the *equivalence classes* defined by R.

Given any element x in S consider all the elements y such that xRy. These form a subset of S: let us call it A_x. We show first that two elements are in the same subset A_x if and only if they stand in the relation R to one another. Suppose yRz and $y \in A_x$. Then xRy and so xRz since R is transitive. Hence $z \in A_x$. Conversely if y and z are both in A_x we have xRy and xRz, i.e. yRx by the symmetric property and xRz: thus yRz by transitivity.

For each element x we now have a subset A_x, but these will not all be distinct. We will show that two such are either mutually exclusive or else identical. Suppose A_x and A_y both have an element z. Then xRz and yRz, and so zRy by the symmetric property; hence xRy by the transitive property. Now take any element w of A_x. Then xRw and since xRy we have yRx, giving yRw. So w is in A_y. Hence we have shown that if A_x and A_y have one element z in common, any element w of A_x is in A_y, i.e. $A_x \subseteq A_y$. Similarly $A_y \subseteq A_x$ and so the subsets A_x and A_y are identical.

Thus we have mutually exclusive subsets $A_{x_1}, A_{x_2}, \ldots$. Finally, by the reflexive property any element t is in one of the subsets, viz. A_t.

Hence S is divided into a set of mutually exclusive subsets as required.

Note that the subset A_x may equally well be described as A_y for any element y in it—the important things are the equivalence classes and not the individual elements.

The equivalence classes theorem may not at first sight seem very exciting, but it is in fact extremely important, though essentially trivial in its proof. It gives us a simple set of necessary and sufficient conditions for a given relation to break up a set into subsets and is most useful, of course, for complicated types of relations. Several important examples of its use will be encountered later.

Examples of relations

1. In the set of all living human beings let xRy mean that x and y are the same age. This is clearly reflexive, symmetric and transitive and so is an equivalence relation. The equivalence classes are those in Example 1 on page 17.

2. In the set of all integers let xRy mean that $x-y$ is divisible by a fixed integer n. This relation is clearly reflexive since 0 is divisible by n, it is symmetric since if $x-y$ is divisible by n so is $y-x$, and it is transitive since if $x-y$ and $y-z$ are both divisible by n so is their sum, which is $x-z$. Hence the relation is an equivalence relation and the equivalence classes are those in Example 2 on p. 17.

3. In the set of all male humans let xRy stand for the fact that x and y have the same parents. This gives an equivalence relation, dividing the human males into equivalence classes of families of brothers.

4. In the set of human males let xRy mean that x is a direct ancestor of y. Although this is transitive it is neither reflexive nor symmetric and so is not an equivalence relation.

5. In the set of *all* humans let xRy mean that x is a first cousin of y. This is symmetric but not transitive or reflexive. Thus the relation is *not* an equivalence relation and does not divide all humans into classes of cousins.

6. In the set of all humans let xRy mean that x and y have at least one parent in common. This is reflexive and symmetric but not transitive and is not an equivalence relation.

Worked exercises

1. Prove that $A \cap \bar{A} = \varnothing$.

Suppose the element x is in both A and in $\bar{A}$. Then since $x \in \bar{A}$ it is not in A by definition of $\bar{A}$, and so we have a contradiction. Hence there is no element common to A and $\bar{A}$, and so $A \cap \bar{A} = \varnothing$.

2. Is the following statement true: $A \supset B \Leftrightarrow A \cap \bar{B} \neq \varnothing$?

If $A \supset B$ then $\exists x.\ x \in A$ but $x \notin B$. Hence $x \in \bar{B}$ and so $A \cap \bar{B} \neq \varnothing$. Hence $A \supset B \Rightarrow A \cap \bar{B} \neq \varnothing$.

Conversely suppose $A \cap \bar{B} \neq \varnothing$. Then $\exists x . x \in A$ but $x \notin B$. But we cannot deduce from this that $A \supset B$, and we give a counter example to disprove this implication.

Suppose A consists of all even integers, and B of all integers divisible by 3. Then 2 is in A but not in B and so $2 \in \bar{B}$, showing that $A \cap \bar{B} \neq \varnothing$. But of course it is not true that $A \supset B$, since $3 \in B$ but not $\notin A$.

A further counter example could be obtained by taking A and B to be any two disjoint sets.

3. In the set of integers are the following equivalence relations?

(*a*) aRb means that $a-b$ is even,

(*b*) aRb means that $a-b$ is odd.

(*a*) $a-a = 0$ is even, and so R is reflexive. If $a-b$ is even, so is $b-a$ and so R is symmetric. If $a-b$ and $b-c$ are both even, so is their sum, $a-c$, and so R is transitive. Hence R is an equivalence relation. R gives us two equivalence classes, one consisting of all even integers and the other of all odd integers.

(*b*) $a-a = 0$ and is not odd, so R is not reflexive. If $a-b$ is odd so is $b-a$, and so R is symmetric. If $a-b$ and $b-c$ are both odd, their sum, $a-c$, is even, and so R is not transitive. Hence R is not an equivalence relation.

Exercises 2A

Are nos. **1–10** sets? If so, give a typical element, and if not explain why not.

1. All the light given out by all the stars at any given moment.

2. All the light waves which are in the process of being given out by all the stars at any given moment.

3. All animals that have ever lived.

4. All animals that have ever lived on the earth.

5. All mammals that have ever lived on the earth.

6. All mammals that are living at the present moment on the earth.

7. All men who are over 200 years old on the earth at the present time.

8. All objects that have been worn at some time on the foot of a human being who is living at present on the earth.

9. Red, green and blue.

10. All colours.

Prove by translating the notation into everyday language, the statements **11–25**, where A, B, ... are sets and a, b, x, ... are elements.

11. $A = B$ and $a \in A \Rightarrow a \in B$.

12. $B \not\ni a$ and $A = B \Rightarrow a \notin A$.

13. $a \in A$ and $A \subseteq B \Rightarrow a \in B$.

14. $a \notin B$ and $A \subseteq B \Rightarrow a \notin A$.

15. $a \in A$ and $a \notin B \Rightarrow A \nsubseteq B$.

16. $A \supset B$ and $b \notin A \Rightarrow b \notin B$.

17. $A \supseteq B$ and $a \in A$, $a \notin B \Rightarrow A \supset B$.

18. $A \subseteq B \Rightarrow A \cup B = B$, $A \cap B = A$.

19. $a \in A \cup B$ and $a \notin A \Rightarrow a \in B$.

20. $a \in A$ and $a \in B \Rightarrow A \cap B \neq \emptyset$.

21. $A \cup B = B \cup A, A \cap B = B \cap A.$

22. $a \in \bigcap_{i=1}^{n} A_i \Rightarrow a \in A_{n-1}.$

23. $A - B \neq \emptyset \Leftrightarrow \exists\, a . a \in A$ and $a \notin B.$

24. $C(A) = \{x : x \notin A\}.$

25. $A = \{x : x \in A\}.$

Are statements **26–34** true? If so prove them, while if not give a concrete counter example.

26. $A \not\subset B \Leftarrow A = B.$

27. $A \not\subset B \Rightarrow A \supseteq B.$

28. $A \subset B$ and $B \subseteq C \Rightarrow A \subset C.$

29. $A \subset B$ and $B \supset C \Rightarrow A = C.$

30. $A \subset B \Rightarrow B - A \neq \emptyset; \quad A \subseteq B \Rightarrow A - B = \emptyset.$

31. $A \subset B \Leftarrow B - A \neq \emptyset; \quad A \subseteq B \Leftarrow A - B = \emptyset.$

32. $A - \bar{B} = \emptyset \Leftrightarrow A \cap B = \emptyset.$

33. $A = \{x_1, x_2, x_4\} \Rightarrow \exists\, r, r \leqslant 4 . x_r \notin A.$

34. $A = \{x_1, x_2, x_4\} \Leftrightarrow x_3 \notin A.$

Translate **35–39** into everyday language.

35. $\forall\, a > 0 \,\exists\, b > 0 . b < a.$

36. $\forall h > 0 \,\exists\, N.\ |u_n - l| < h \,\forall\, n \geqslant N.$

37. $(A \cup B) \cap C \neq \emptyset.$

38. $\exists\, a \in (A - \bar{B}).\ |a - l| < \delta.$

39. $[\{a, b, c\} \cup \{a, b, d\}] \cap \{d, e\} \neq \emptyset.$

40. If A is the set of all dogs, B is all those with black tails, C all those with white heads, D all female dogs and E all dogs with black tails and black heads, interpret the following:

(i) $B \cup C$; (ii) $B \cap C$; (iii) $A - D$; (iv) $C \cup E$;
(v) $C \cap E$; (vi) $A \cap \bar{E}$; (vii) $A \cap B \cap E$; (viii) $\bar{D} \cup E$.

41. If A is the set of all positive integers, B all even positive integers, C all positive integers divisible by 3, D all positive integers divisible by 4 and E all primes, interpret the following:

(i) $A - B$; (ii) $B \cap C$; (iii) $B - D$; (iv) $C \cap E$;
(v) $C \cup E$; (vi) $A \cap \bar{C}$; (vii) $E \cap \bar{B}$.

Are the relations in **42–50** equivalence relations? If not, state which laws are not satisfied; if they are give the equivalence classes.

42. In the set of real numbers aRb means $a \leqslant b$.

43. In the set of real numbers aRb means $a - b$ is an integer.

44. In the set of positive integers aRb means a and b are co-prime.

45. In the set of all humans, aRb means a is a friend of b.

46. In the set of all living humans aRb means a and b live in the same house.

47. In the set of all humans living in England, aRb means a and b live in the same county.

48. In the set of all triangles in a plane aRb means a and b are congruent.

49. In the set of all circles in a plane aRb means a and b meet in at least one real point.

50. In the set of all lines in a plane aRb means a and b meet at the origin.

Exercises 2B

Prove **1–5**, where $A, B, \dots$ are sets and $a, b, x, \dots$ are elements.

1. $A \subseteq B \Leftrightarrow \forall x . x \notin A, x \in B.$

2. $\bar{A} \subset B \Rightarrow \exists x . x \in A \cap B.$ Is the converse true?

3. $A \cap B \subseteq A \cup B;\ A \cap B = A \cup B \Leftrightarrow A = B.$

4. $(A \cup B) - (A \cap B) = \{x : x \in$ either A or B but not both$\}$.

5. $\bar{\bar{A}} = A;\ A - (A - B) = A \cap B.$

6. Give an example of an infinite ascending string of sets

$$A_1 \subseteq A_2 \subseteq A_3 \subseteq \dots.$$

7. Show that if aRa and also if aRb and $bRc \Rightarrow cRa$ then R is an equivalence relation.

8. Let P be a relation in a set S which is reflexive and transitive but not necessarily symmetric. Define a second relation R such that aRb if and only if *aPb and bPa*. Show that R is an equivalence relation. Illustrate by a concrete example.

9. In the set of points (x, y) in a plane let $(a, b) R (c, d)$ mean that $a+d = b+c$. Prove that R is an equivalence relation and find its equivalence classes.

10. In the set of pairs (m, n) where m, n are positive, negative or zero integers but $n \neq 0$, define $(a, b) R (c, d)$ to mean that $ad = bc$. Show that R is an equivalence relation and find its equivalence classes.

3

ELEMENTARY SET THEORY

3.1. Abstract sets

The idea of a set is so general that it might be thought impossible to develop any theory of sets as such. It is true that in most mathematics we work within one particular set or type of set and that most of our results are valid for that only, but there are also many theorems which apply to all sets. An example will help to make this clear.

The theorem that any number can be factorised as a product of primes in one and only one way presupposes that we are working within the set of integers, by its very wording. With a suitable definition of factorisation and of prime it may be extended to certain other sets, but it cannot be true in any sense of a completely arbitrary set: it would be difficult to give it a meaning when applied to the set of all cows, for instance.

The theorem that $\overline{A \cup B} = \bar{A} \cap \bar{B}$, on the other hand, is of a different type. It applies to any sets A and B whatsoever. For example, if A is the set of all even integers and B that of all odd ones, $\overline{A \cup B}$ is the set of all objects that are not integers, and all these are neither even nor odd and so are in $\bar{A} \cap \bar{B}$ also, and conversely. The same reasoning applies if A is the set of all cows and B that of all animals more than 3 years old.

There are many theorems like the preceding example which apply to sets in general. The study of them is that branch of mathematics known as 'Set theory' which, although fairly recent in its development, has been studied intensively and fruitfully. Its results of course have an extremely wide application and, at any rate in its elementary parts, it is fairly easy to understand. In this chapter we will study a very little of the theory, our approach being intuitive rather than rigorous. A more abstract approach is left until volume 2.

Because our work will be so general we are not interested in particular concrete identifications of the sets used—A, B, etc., will stand for *any* sets. Considering individual cases is not very fruitful in this particular subject, but the reader may easily construct them for himself if he wishes, and we will occasionally give a few. The application of any theorem to several examples and the investigation of its meaning there will soon convince the doubtful student of its universal validity. The important point is that we are considering sets as abstract entities and studying properties that they have in common because they are sets; it is analogous to elementary arithmetic, where we consider numbers as abstract objects and study their properties as such, ignoring the fact of whether they stand for apples, or cakes, or grains of sand. In both cases any results that we may discover may then be applied to any particular cases.

The work that we do must involve only those ideas which are common to all sets—such as union, intersection, complement and the other notions introduced in §2.2. It may all seem very obvious, partly because there are so few basic ideas with which we can start, partly because we are dealing only with the elementary parts of the subject, but the reader should find it interesting and may discover new vistas opened to him, especially if he is accustomed to thinking of mathematics merely as dealing, explicitly or implicitly, with numbers.

3.2. Methods of proof in set theory

There are several methods of attacking elementary problems of set theory, and in this section we will consider three of them, and mention a fourth to be dealt with in the following section; showing their application to three typical problems. The problems to be considered are:

(*a*) Prove that $A \cup \overline{B} = \overline{B-A}$.

(*b*) Show that $A \subseteq \overline{B} \cap \overline{C} \Rightarrow A \cap B \cap C = \varnothing$ and discover whether or not the converse implication is true.

(*c*) Simplify the set $\overline{\overline{A} \cap \overline{B}}$.

***Method 1**. By consideration of individual elements.*

(*a*) Suppose the element x is in $A \cup \overline{B}$. Then either $x \in \overline{B}$ or $x \in A$, so that if $x \in B$ it must be in A, i.e. it cannot be in $B-A$. Hence $x \in \overline{B-A}$, and so $A \cup \overline{B} \subseteq \overline{B-A}$.

Conversely suppose $x \in \overline{B-A}$. Then $x \notin B-A$, so that if $x \in B$ it is also in A, and hence is either in A or not in B, i.e. $x \in A \cup \bar{B}$. Thus $A \cup \bar{B} \supseteq \overline{B-A}$. Hence $A \cup \bar{B} = \overline{B-A}$.

(*b*) Suppose $A \subseteq \bar{B} \cap \bar{C}$. Then any element x of A is in both $\bar{B}$ and $\bar{C}$, i.e. it is in neither B nor C, and so certainly not in $B \cap C$. Hence $A \cap (B \cap C) = \varnothing$.

Conversely, if $A \cap B \cap C = \varnothing$ no element x of A can be in *both* B and C, but it may of course be in one of them. Thus x need not be in *both* $\bar{B}$ and $\bar{C}$, and so the implication that $A \cap B \cap C = \varnothing \Rightarrow A \subseteq \bar{B} \cap \bar{C}$ cannot be proved, at any rate by our argument.

To disprove the implication properly we need a counter example, and it is left to the reader to furnish one. (Find one in which $A \cap B \neq \varnothing$.)

(*c*) If $x \in \overline{\bar{A} \cap \bar{B}}$ then $x \notin \bar{A} \cap \bar{B}$. Hence x is not in both $\bar{A}$ and $\bar{B}$, and so is in one at least of A and B, i.e. is in $A \cup B$. Hence $\overline{\bar{A} \cap \bar{B}} = A \cup B$. (Notice that the argument works both ways, in an if and only if manner, and so the two sets $\overline{\bar{A} \cap \bar{B}}$ and $A \cup B$ are equal: we do not merely have one contained in the other.)

This method is the most natural one, and the arguments are largely a matter of common sense. It has the defect that in complicated cases it becomes difficult to follow the logic without some diagrammatic aid.

Method 2. *The truth table method.*

(*a*) We consider all possible types of elements, those in A and B, those in A but not B, those in B but not A and those in neither.

Suppose $x \in A$ and B. Then $x \in A \cup \bar{B}$. Also $x \notin B-A$ and so $x \in \overline{B-A}$.

Suppose $x \in A$ but $\notin B$. Then $x \in A \cup \bar{B}$. Also $x \notin B-A$ and so is in $\overline{B-A}$.

If $x \notin A$ but $\in B$ it is not in $\bar{B}$ and so $x \notin A \cup \bar{B}$. Also $x \in B-A$ and so is not in $\overline{B-A}$.

Finally let $x \notin A$ and $\notin B$. Then $x \in \bar{B}$ and so is in $A \cup \bar{B}$. Also $x \notin B-A$ and so is in $\overline{B-A}$.

Thus in all four cases x is either in both or in neither of the sets $A \cup \overline{B}$ and $\overline{B-A}$, and thus these two sets are equal, since the four cases take care of all possible objects x.

(*c*) If $x \in A$ and B it is not in $\overline{A} \cap \overline{B}$ and so *is* in $\overline{\overline{A} \cap \overline{B}}$. Similarly if x is in one of A and B. If x is in neither A nor B it is in $\overline{A} \cap \overline{B}$ and so is not in $\overline{\overline{A} \cap \overline{B}}$. Hence $\overline{\overline{A} \cap \overline{B}}$ consists of all elements in at least one of A and B, and so $\overline{\overline{A} \cap \overline{B}} = A \cup B$.

The method can in certain cases be quite powerful and easy to apply. If there are many sets involved, however, it becomes unwieldy, for if we have n sets we have 2^n cases to consider, although some may often be taken together. In some examples, such as (*b*), it is not a convenient method to use.

Method 3. *By use of theorems.*

In volume 2 we will meet quite a number of postulates and theorems of set theory, and these may be used to prove other results. This is perhaps the best method for complicated problems (sometimes it will be the only feasible one) but it is not necessary for the simple problems that we meet at present, and we do not in fact wish to introduce all the necessary theorems at this stage. We illustrate it for examples (*b*) and (*c*), combining it with intuitive basic logic: used in this way it has the merit of directness and does not necessitate the introduction of individual elements.

(*b*) It is an easily proved theorem that $\overline{B} \cap \overline{C} = \overline{B \cup C}$. Hence if $A \subseteq \overline{B} \cap \overline{C}$ we have $A \subseteq \overline{B \cup C}$ and so $A \cap (B \cup C) = \varnothing$. Hence certainly $A \cap B \cap C = \varnothing$.

Conversely if $A \cap B \cap C = \varnothing$ we certainly have $A \cap (B \cap C) = \varnothing$ but not necessarily $A \cap (B \cup C) = \varnothing$. Hence we cannot say that $A \subseteq \overline{B \cup C}$, i.e. that $A \subseteq \overline{B} \cap \overline{C}$, and we can easily find a counter example as with method 1.

(*c*) $\overline{A} \cap \overline{B} = \overline{A \cup B}$ and so $\overline{\overline{A} \cap \overline{B}} = \overline{\overline{A \cup B}} = A \cup B$.

Method 4. *By use of Venn diagrams.*

The best intuitive method, and the easiest to follow, is by means of certain diagrams known as Venn diagrams. Since this method appeals to concrete figures, and therefore to one special representation of a set, it is not absolutely rigorous, but it is quite adequate for our purposes, and it is recommended that the beginner use it at least to illustrate his problems, though he should also become competent in the use of method 1. Venn diagrams are described in the following section.

3.3. Venn diagrams

In a Venn diagram we represent each basic set with which we are dealing by the inside of some simple geometrical shape drawn on our paper—circles are usually used. Sets which are formed from our basic ones, by union or intersection or complement,

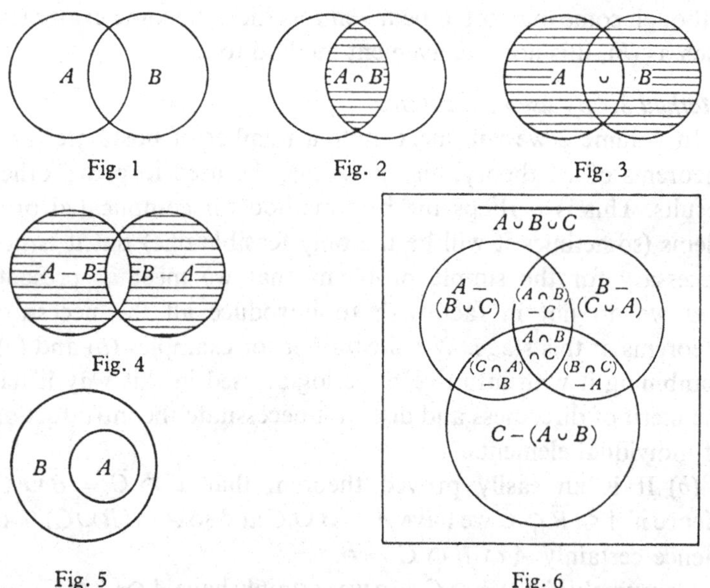

Fig. 1 Fig. 2 Fig. 3 Fig. 4 Fig. 5 Fig. 6

are then represented in an obvious way by the union, intersection or complement of the shapes. For example, suppose we have two basic sets A and B. We represent them by circles as in figure 1, and then $A \cap B$ is represented by the part common to the two circles, shaded in figure 2, while $A \cup B$ is the part shaded in figure 3. We may similarly interpret $A - B$ and $B - A$ as in figure 4.

The universal set is often, though not always, represented by a rectangle enclosing all the other sets. The complement of A is then the space inside the rectangle that is outside A. In practice the rectangle is often omitted, the universal set being taken to be the whole sheet of paper, and in this case the complement of A is merely that part of the paper outside A.

The fact that $A \subset B$ is represented by the circle A being placed inside B, as in figure 5.

To give a further example we show in figure 6 a Venn diagram for three sets A, B and C, and show what each area of the diagram represents.

The advantage of Venn diagrams is that they give a simple and natural picture of the sets with which we are dealing, without mentioning the individual elements. It is usually easy to solve problems by their use, and we proceed to give some examples of this, starting with those of §3.2.

(*a*) Representing A and B on a Venn diagram as in figure 7 we see that both $A \cup \overline{B}$ and $\overline{B-A}$ are represented by the shaded area, and so are equal.

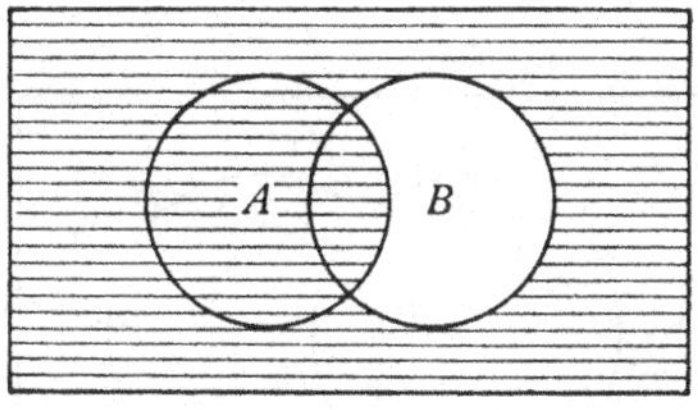

Fig. 7

(*b*) We first represent B and C by circles. (Note that we must draw them to intersect since we have no knowledge as to their relative positions and so must take them in general position, which makes provision for intersection—if in fact $B \cap C = \emptyset$ this would merely mean that no elements were 'inside' their common part, speaking loosely.) If $A \subseteq \overline{B} \cap \overline{C}$ we must have A lying completely outside both B and C and draw it accordingly, as in figure 8. We now see immediately from the diagram that $A \cap B \cap C = \emptyset$.

If on the other hand we know that $A \cap B \cap C = \emptyset$ it does not necessarily follow that A lies outside both B and C, since we can deduce only that the three circles have no common part —we may have a diagram as in figure 9.

Note that we must still use a little basic logic to interpret our conditions in order to draw the sets in the correct respective positions; once this is done the result is more or less obvious.

(*c*) $\bar{A} \cap \bar{B}$ is the part of the diagram outside both A and B, and so its complement consists merely of $A \cup B$.

The above examples are of properties common to all sets, but Venn diagrams are also used to illustrate particular cases and make deductions from them. Examples are given below.

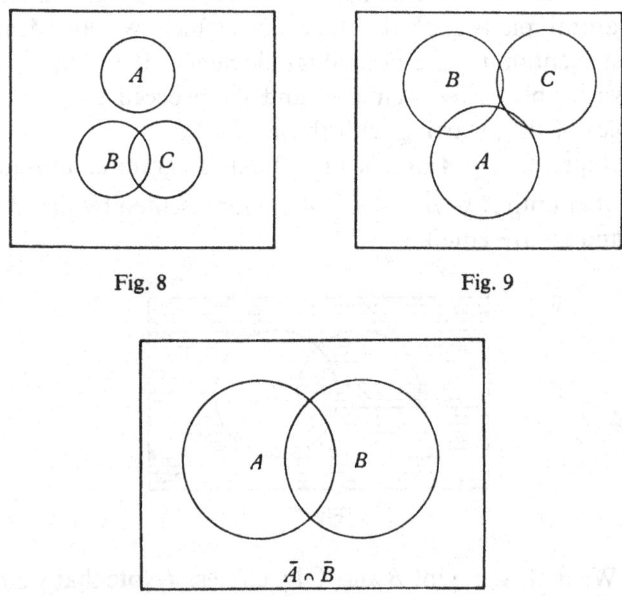

Fig. 8

Fig. 9

Fig. 10

Example 1. We consider various types of rabbit. We represent first the set of all rabbits with brown heads. This includes all those which are completely brown, so this latter circle is placed within the former. The set of all those with white tails may overlap that of the brown-headed ones, but not the set of brown ones, while the all-white rabbits are inside the white-tailed ones and do not include any of the brown-headed ones. Finally, all Belgian hares are completely brown and all my rabbits have brown heads and white tails. The situation is shown in figure 11. From the Venn diagram we may deduce various things. For instance, none of my rabbits can be Belgian hares, nor can any Belgian hares have white tails. A rabbit with a brown head may have a white tail or it may not, though if it is mine it must have.

None of these properties of rabbits is very startling: they all follow easily from the very language used. The Venn diagram,

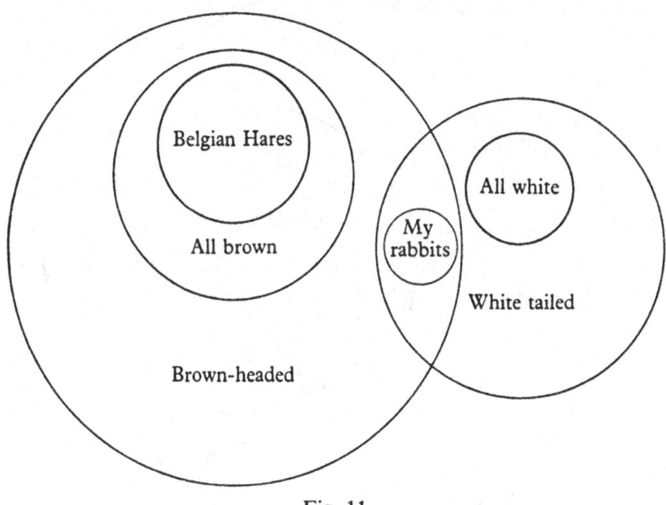

Fig. 11

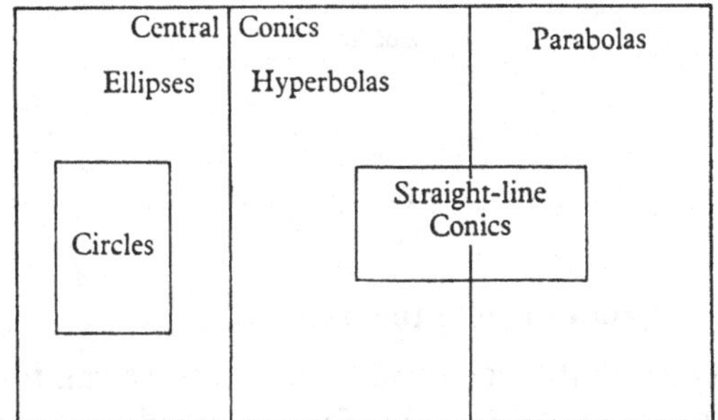

Fig. 12

however, expresses them in simple visual terms and enables us to see at a glance the various interactions of the properties.

Example 2. Figure 12 shows a Venn diagram for the different types of conics. From it we may see quickly the various possibilities and their mutual inclusions and exclusions, for example, that no circles are also parabolas, or that straight line conics may be either central or not.

Example 3. Figure 13 shows various types of humans, categorised by the relationships that they possess.

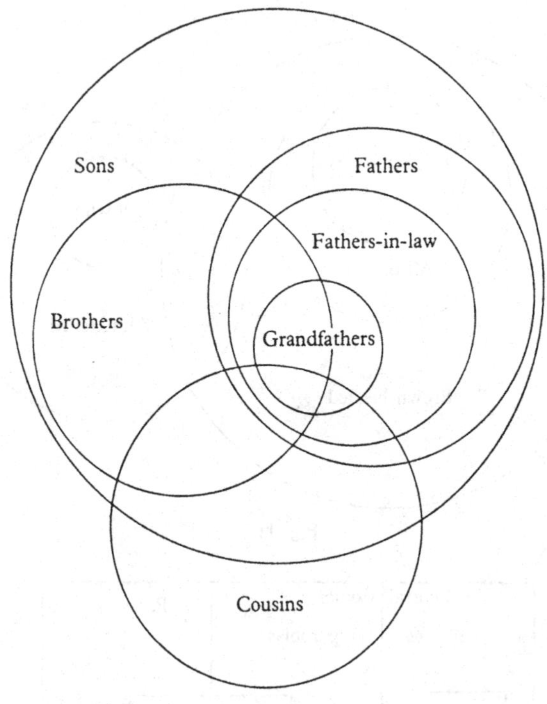

Fig. 13

3.4. Simple algebra of sets: the basic rules

We have seen that given certain basic sets we can form others from them by using the notions of union, intersection, difference and complement, much in the same way that we can form new numbers from given ones by the processes of addition, subtraction, multiplication and division. In the latter case we manipulate our numbers according to certain rules, which we learn in elementary arithmetic and algebra. These depend ultimately on a very few basic laws—the *fundamental laws of algebra*—which are, briefly:

The Commutative Laws. $x+y = y+x$ and $xy = yx$.

The Associative Laws. $x+(y+z) = (x+y)+z$ and $x(yz) = (xy)z$.

The Distributive Law. $x(y+z) = xy+xz$.

The Zero and Unity Laws. $x+0 = 0+x = x$ and $1.x = x.1 = x$.

The Subtraction and Division Laws. $x+(-x) = (-x)+x = 0$ and $x(1/x) = (1/x)x = 1$.

A full discussion of these is given in chapter 10: the important point at present is to notice that they form the basis of ordinary arithmetic and algebra. In the same way the manipulation of sets by use of union, intersection, etc., depends ultimately on similar basic rules.

We proceed to give some of the fundamental laws of set theory. They are all very easy to demonstrate, either by use of Venn diagrams or by consideration of elements, and the verification will sometimes be left to the reader. They are consequences of the so-called 'Laws of Logic', though in a rigorous development of set theory they would often be taken as postulates. This type of approach is taken in volume 2, but at the present we are interested merely in seeing the truth of the rules. We also ignore the fact that those we give are not all independent: some may be deduced directly from the others.

The reader is not advised to learn the rules by heart at this stage. It is usually easier to solve elementary set problems directly rather than by use of laws—these are more necessary in advanced work. What must be noted is the close similarity between certain of these rules for sets and the fundamental laws of algebra. We tend to think of algebra as being concerned with numbers, which may be combined according to the rules of addition, subtraction, multiplication and division. We will find in later chapters that these four processes are not restricted to numbers, while in the present chapter we are meeting completely different processes which nevertheless obey some, but not all, of the same fundamental laws. The fact of the existence of this analogy is vital to the understanding of abstract algebra, and we will carefully point out how far the similarity goes. We give the rules partly for their innate interest and partly in order to point out the analogy with elementary algebra. Although the ideas with which we deal are important throughout the book, the detailed algebra of sets is not used in most of the succeeding chapters.

The laws for inclusion

1 *The Reflexive Law.* $A \subseteq A$, where A is any set.

2 *The Anti-symmetric Law.* $A \subseteq B$ and $B \subseteq A \Rightarrow A = B$.

3 *The Transitive Law.* $A \subseteq B$ and $B \subseteq C \Rightarrow A \subseteq C$.

Note the similarity between these and the laws for numbers that $x \leqslant x$, $x \leqslant y$ and $y \leqslant x \Rightarrow x = y$, $x \leqslant y$ and $y \leqslant z \Rightarrow x \leqslant z$. We see that to a certain extent sets have a magnitude, or rather an order. The great difference is that only certain pairs of sets may be compared in this way: given two arbitrary sets A and B we cannot say that either $A \subseteq B$, $A = B$ or $A \supseteq B$. With numbers this can be said, the consequence being that all real numbers may be represented on a straight line: they are 'totally ordered'. Sets are 'partially ordered', while complex numbers are not ordered at all.

The laws for union and intersection

4 *The Commutative Laws.* (*a*) $A \cup B = B \cup A$;
(*b*) $A \cap B = B \cap A$.

5 *The Associative Laws.* (*a*) $A \cup (B \cup C) = (A \cup B) \cup C$;
(*b*) $A \cap (B \cap C) = (A \cap B) \cap C$.

6 *The Distributive Laws.* (*a*) $A \cup (B \cap C) = (A \cup B) \cap (A \cup C)$;
(*b*) $A \cap (B \cup C) = (A \cap B) \cup (A \cap C)$.

7 *The Idempotent Laws.* (*a*) $A \cup A = A$;
(*b*) $A \cap A = A$.

8 *The Absorption Laws.* (*a*) $A \cup (A \cap B) = A$;
(*b*) $A \cap (A \cup B) = A$.

The first thing to notice about these laws is the connection between the '*a*' group and '*b*' group. To form a '*b*' from the corresponding '*a*' we merely interchange the signs $\cup$ and $\cap$.

In other words there is no difference *within the laws* between union and intersection: if different symbols were used we could not tell from the laws which was union and which intersection. This duality carries through all our set theory: it shows us that algebraically the processes $\cup$ and $\cap$ behave exactly the same, although of course both sometimes enter into one piece of work as in the Distributive and Absorption Laws. Whenever we prove a theorem we may at once write down the corresponding one obtained by interchanging $\cup$ and $\cap$, just as in two-dimensional projective geometry we have a duality between points and lines.

The Commutative Laws state that it does not matter in which order we take the two sets which are to be combined: this may seem an obvious remark and so it is for this case, but it needs stating, since we will later meet examples of processes which are not commutative (e.g. in §5.5). The Associative Laws again seem fairly obvious: in forming the union or intersection of three sets it does not matter which pair we combine first. As a result we may write $A \cup B \cup C$ or $A \cap B \cap C$ without ambiguity. As a result of the Commutative and Associative Laws we may combine any number of sets in any order, and from the Idempotent Laws we may also ignore any repeats of the same set. For example, $(A \cup B) \cup (A \cup C) = A \cup B \cup C$.

Laws 4 and 5 are analogous to those for ordinary algebra, and so is $6b$ if we replace $\cup$ by sum and $\cap$ by product, *but not vice versa*, i.e. $x(y+z) = xy+xz$, but it is not true that $x+yz = (x+y)(x+z)$. We have here an important difference between our set algebra and elementary algebra: the latter is not symmetrical in $+$ and $\times$, so that one Distributive Law is true while the other is not.

The laws 7 and 8 have no analogues in elementary algebra, so that while we may combine any number of added terms in any order we may *not* ignore the repeated ones: $(2+3)+2 \neq 2+3$.

The proof of 4 and 7 is immediate, and that of the others is straightforward. We give in figure 14 Venn diagrams to illustrate them and also prove $6b$ by elements.

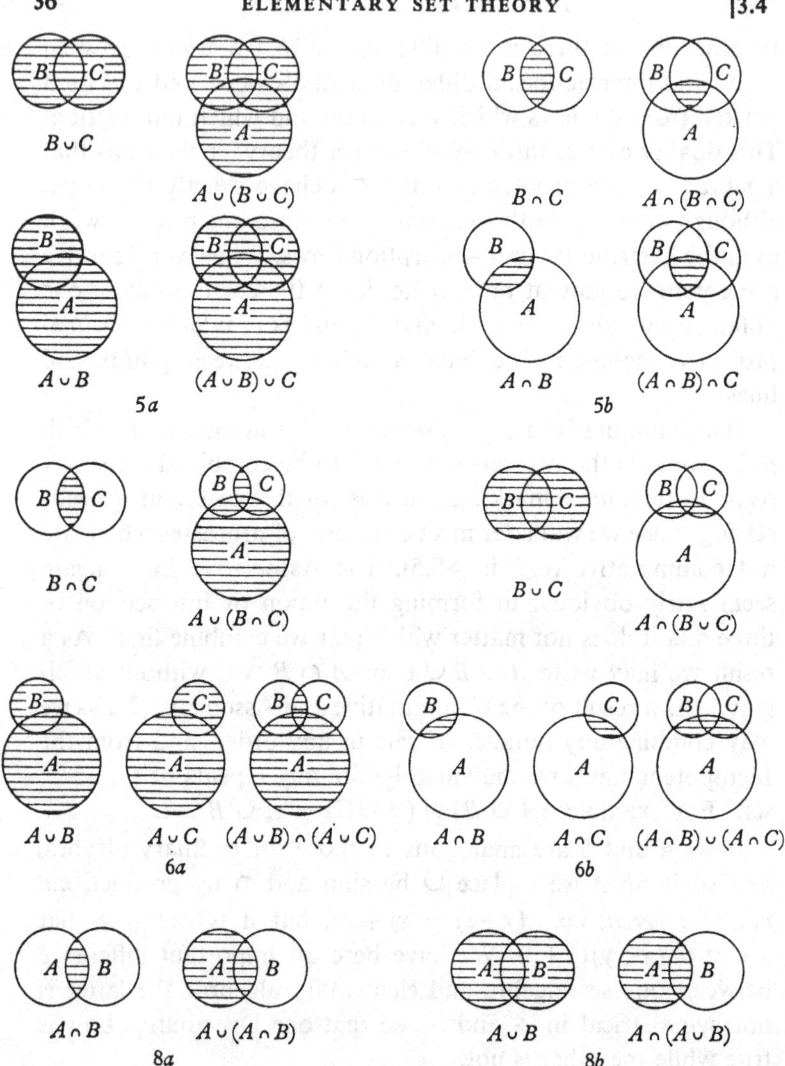

Fig. 14

Proof of 6b by elements

Suppose $x \in A \cap (B \cup C)$. Then $x \in A$ and $x \in$ one at least of B and C. Thus x is either in both A and B or in both A and C, or both, i.e. $x \in (A \cap B) \cup (A \cap C)$. Thus

$$A \cap (B \cup C) \subseteq (A \cap B) \cup (A \cap C).$$

Conversely suppose $x \in (A \cap B) \cup (A \cap C)$. Then either

$x \in A \cap B$, or $x \in A \cap C$, or both, i.e. $x \in A$ and $x \in$ at least one of B and C, and so $x \in A \cap (B \cup C)$. Thus

$$A \cap (B \cup C) \supseteq (A \cap B) \cup (A \cap C).$$

The laws for the empty set and the universal set

The empty set is as usual denoted by ø, and for the present we will write I to mean the universal set. Remember that I means the set of all elements which enter into the particular piece of work with which we are concerned.

$$9a,\ A \cup ø = A; \qquad 10a,\ A \cup I = I;$$

$$9b,\ A \cap I = A. \qquad 10b,\ A \cap ø = ø.$$

Note again the duality, and that to form the dual laws we must interchange ø and I, as well as $\cup$ and $\cap$.

Laws 9a and 9b show that ø and I behave like 0 and 1 in elementary algebra. Thus adding 0 or multiplying by 1 does not change an ordinary number, while the union with ø or intersection with I does not change a set. Law 10b is analogous to multiplication by 0, but 10a has no analogue.

The laws for complements

$$11a,\ A \cup \bar{A} = I; \qquad 12,\ \bar{\bar{A}} = A. \qquad 13a,\ \bar{ø} = I;$$

$$11b,\ A \cap \bar{A} = ø. \qquad 13b,\ \bar{I} = ø.$$

The De Morgan laws

$$14a,\ \overline{A \cup B} = \bar{A} \cap \bar{B}; \quad 14b,\ \overline{A \cap B} = \bar{A} \cup \bar{B}.$$

In law 9 we saw that ø and I behave like 0 and 1, with union and intersection taking the place of + and ×. Let us see if we can find an analogue of $-x$ and $1/x$. $-x$ has the property that $x+(-x) = 0$: its analogue must be such that $A \cup B = ø$. But this is soon seen to be impossible if $A \neq ø$, since for any set B, $A \cup B$ contains A and so cannot be empty. Similarly, an analogue of $1/x$ would have the property that $A \cap B = I$ which again is impossible unless $A = I$. Notice that laws 11a and 11b give both analogues to be $\bar{A}$ at first sight, but these are not true ones since the roles of ø and I, laid down in connection with $\cup$ and $\cap$ in law 9, have been interchanged.

The De Morgan laws are interesting, and the reader should work through the proof. We give a Venn diagram for the first of them.

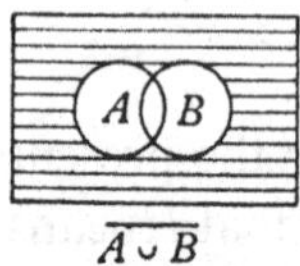

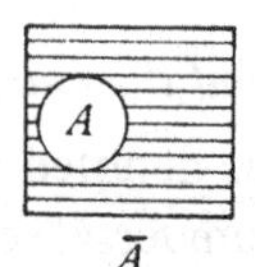

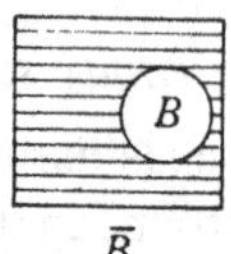

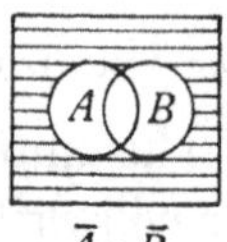

Fig. 15

3.5. Simple algebra of sets: some theorems

Theorem 3.5.1. *The generalised Distributive Laws.*

$$A \cup (B \cap C \cap D) = (A \cup B) \cap (A \cup C) \cap (A \cup D)$$

$$(A \cap B) \cup (C \cap D) = (A \cup C) \cap (A \cup D) \cap (B \cup C) \cap (B \cup D)$$

and so on, as in elementary algebra.

The generalisation is fairly obvious, but it should be noticed that it can be proved using only the Distributive, Associative and Commutative Laws, as follows:

$A \cup (B \cap C \cap D)$

$= A \cup (B \cap (C \cap D))$ assuming the Associative Law

$= (A \cup B) \cap (A \cup (C \cap D))$ by the Distributive Law $6a$.

$= (A \cup B) \cap (A \cup C) \cap (A \cup D)$ again by $6a$ and assuming 5.

$(A \cap B) \cup (C \cap D)$

$= ((A \cap B) \cup C) \cap ((A \cap B) \cup D)$ by $6a$.

$= (C \cup (A \cap B)) \cap (D \cup (A \cap B))$ by $4a$.

$= (C \cup A) \cap (C \cup B) \cap (D \cup A) \cap (D \cup B)$ by $6a$ and assuming the Associative Laws.

$= (A \cup C) \cap (A \cup D) \cap (B \cup C) \cap (B \cup D)$ by $4a$ and $4b$.

Theorem 3.5.2. *The uniqueness of ø and I.*

ø and I are the only sets with the properties given in 9a and 9b, respectively, i.e. if B is a set such that $A \cup B = A$ for all A, then $B = ø$, and if $A \cap B = A$ for all A then $B = I$.

Suppose $A \cup B = A \;\forall\; A$. Then $\emptyset \cup B = \emptyset$. But $\emptyset \cup B = B$ by $9a$, and so $B = \emptyset$.

Suppose $A \cap B = A \;\forall\; A$. Then $I \cap B = I$, but $I \cap B = B$ by $9b$ and so $B = I$.

Note 1. The proof of the second part is dual to that of the first: we interchange $\cup$ and $\cap$, together with $\emptyset$ and I. There was strictly speaking no need to prove the second part separately: we could have stated its correctness by this principle of duality.

Note 2. Although $\emptyset$ is the only set such that $A \cup \emptyset = A$ for all A, there are other sets with this property for particular A's, viz. any set B such that $B \subseteq A$.

Note 3. The theorem is obvious intuitively, and many proofs could be constructed. The one we have given is noteworthy because it proves the theorem using only the basic law 9 and the Commutative Laws.

Theorem 3.5.3. *The Cancellation Law.*

If A, X and Y are three sets such that both $A \cup X = A \cup Y$ and $A \cap X = A \cap Y$ then $X = Y$.

The easiest way to see this is to consider a Venn diagram: the parts of X and Y inside A coincide since $A \cap X = A \cap Y$, and the parts outside coincide since $A \cup X = A \cup Y$. A proof from the basic laws is of interest and we give one below. The reader who finds this type of manipulation difficult to follow is advised to omit this proof, and in any case there is no need for it to be memorised.

$$\begin{aligned} X &= X \cap (X \cup A) \quad \text{by law } 8b \\ &= X \cap (Y \cup A) \quad \text{since } X \cup A = Y \cup A \\ &= (X \cap Y) \cup (X \cap A) \quad \text{by } 6b \\ &= (X \cap Y) \cup (Y \cap A) \quad \text{by hypothesis} \\ &= Y \cap (A \cup X) \quad \text{by } 6b \text{ and the Commutative Laws} \\ &= Y \cap (A \cup Y) \quad \text{by hypothesis} \\ &= Y \text{ by } 8b \text{ again.} \end{aligned}$$

Corollary. *If X is a set such that $A \cup X = I$ and $A \cap X = \emptyset$ then $X = \bar{A}$, i.e. $\bar{A}$ is the only set with properties* 11*a and* 11*b.*

Note. If we have one only of the conditions of the hypothesis in 3.5.3, say that $A \cup X = A \cup Y$ we cannot deduce that $X = Y$. This gives a further difference between the processes $\cup$ and $\cap$ and those of $+$ and $\times$, for $a+x = a+y$ *does* imply that $x = y$.

Theorem 3.5.4. *Some laws for the difference $A-B$:*

(i) $A-B = A \cap \bar{B}$;

(ii) $A-(A-B) = A \cap B$;

(iii) $B \cup (A-B) = A \cup B$;

(iv) $B \cap (A-B) = \varnothing$;

(v) $A-(B \cup C) = (A-B) \cap (A-C)$;

(vi) $A-(B \cap C) = (A-B) \cup (A-C)$.

(i) This is easy to verify. It expresses the fact that $A-B$ means the part of $\bar{B}$ which is also in A, and shows us that there is no need for a separate symbol '$-$'. It is much easier in advanced work to manipulate the symbols $\cup$, $\cap$ and complement, and problems involving $-$ are usually treated in this way. The concept of difference, however, remains useful for us at this stage. $A-B$, that part of $\bar{B}$ in A, is often called the *relative complement* of B with respect to A: the relative complement with respect to I being of course merely $\bar{B}$.

(ii) Again easy to verify from a Venn diagram. Notice the result: the right-hand side is not B as in ordinary algebra, but is that part of B which is in A.

(iii) and (iv) are again easy to verify.

(v) and (vi) give a type of Distributive Law, but notice the interchange of $\cup$ and $\cap$. Venn diagrams for these are given in figure 16.

Theorem 3.5.5. $A-B = B-A = \varnothing \Leftrightarrow A = B$.

Given $A = B$ the left-hand side follows at once.

If $A-B = \varnothing$ all elements of A must be in B, i.e. $A \subseteq B$. If $B-A = \varnothing$ we have similarly that $B \subseteq A$. Hence $A = B$.

Theorem 3.5.6. *The consistency theorem.*

The 3 statements (α) $A \subseteq B$,

(β) $A \cap B = A$,

(γ) $A \cup B = B$

are equivalent, i.e. any one implies the others.

Starting from any one of α, β or γ we may easily prove the others. Let us, however, prove all three equivalent by the minimum of effort, by showing that $\alpha \Rightarrow \beta$, $\beta \Rightarrow \gamma$ and $\gamma \Rightarrow \alpha$. We then have, diagrammatically, that

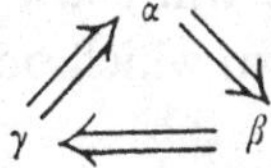

and it can be seen that this is sufficient to prove all 3 equivalent.

$\underline{\alpha \Rightarrow \beta}$. If $A \subseteq B$ all A is in B and so $A \cap B = A$.

$\underline{\beta \Rightarrow \gamma}$. If $A \cap B = A$ then the Absorption Law 8a with A and B interchanged gives $B \cup (B \cap A) = B$, i.e. $B \cup A = B$.

$\underline{\gamma \Rightarrow \alpha}$. If $A \cup B = B$ we must have no part of A outside B, i.e. $A \subseteq B$.

The consistency theorem enables us to replace inclusion signs by either union or intersection, another matter of importance in the advanced theory but not necessary for our present purposes.

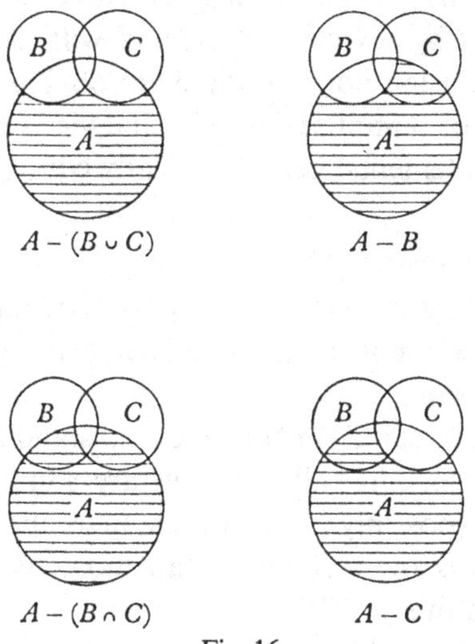

Fig. 16

3.6. Applications to logic: the algebra of statements

It is a remarkable fact that a large part of ordinary logical reasoning can be reduced to an algebra which is much the same as the algebra of sets. This was first done by George Boole

(1815–1864) in his book, *The Laws of Thought*, published in 1854 and forming a great landmark in the history of modern algebra. The subject is known as 'symbolic logic', or sometimes 'Boolean algebra', though the latter term has now been given a rather more precise meaning. We will show how statements may be expressed in a symbolic form which obeys the same rules as set algebra.

Let A, B, C, ... stand for certain statements, which may be either true or false. As an example we will suppose that

$$\left.\begin{array}{l}A\\ B\\ C\\ D\end{array}\right\}\text{ stands for the statement that }\left\{\begin{array}{l}\text{all monkeys are stupid,}\\ \text{apples are good this year,}\\ \text{I like cats,}\\ \text{all mammals are stupid}\end{array}\right.$$

Inclusion

The truth of each of the statements A, B and C above is independent of the others, but if D is true then A is also. This fact is expressed by saying $D \subseteq A$, and with this meaning the laws 1–3 for inclusion hold. 1 and 3 are obvious, while 2 states that if the truth of B implies that of A and vice versa, then A and B are 'equal' statements, i.e. are logically equivalent.

Union and intersection

In logic we need often to connect two statements to form a third, the two common connectives being the words 'or' and 'and' or equivalents.

Let us write $\vee$ to stand for 'or' and $\wedge$ to stand for 'and'. Then $A \vee B$ is the statement: 'all monkeys are stupid *or* apples are good this year' (possibly both), which again may be either true or false. $A \wedge B$ is the statement: 'all monkeys are stupid *and* apples are good this year'.

Let us examine the laws 4–8 with this definition, replacing $\cup$ and $\cap$ in the laws by the above symbols $\vee$ and $\wedge$. The Commutative and Associative Laws need only a moment's thought. For example, the two statements (i) 'all monkeys are stupid or apples are good this year', and (ii) 'apples are good this year or all monkeys are stupid' are equal in the sense of being logically equivalent: the truth of either implies the truth of the other. The Idempotent Laws are also obviously satisfied.

The first Distributive Law says that the two statements (using A, B and C instead of our particular example both to save space and to make the argument general) (i) 'either A is true or both B and C are true', and (ii) 'either A or B is true and also either A or C is true' are equivalent, and a little thought convinces us that this is so. Similarly, for the second Distributive Law.

The first Absorption Law says that A is equivalent to the statement that 'either A or both A and B are true', and this is so since the case where both A and B hold is included in the case where A alone is true. Again the second Absorption Law holds similarly.

ø and I

We take any statement which is necessarily true as I and any which is false as ø. For example, I could be: 'all monkeys are mammals' and ø: 'no apples contain pips'. Then a little thought will convince us of the validity of laws 9 and 10. Notice that I and ø are not unique statements, but that any two of the different statements for either are logically equivalent.

Complements

$\overline{A}$ is taken to mean the negation of A, i.e. in our examples $\overline{A}$ would be: 'not all monkeys are stupid' or 'there are some monkeys that are not stupid' (*not* 'all monkeys are not stupid', which means that no monkeys are stupid), while $\overline{B}$ would be: 'apples are not good this year'.

Laws 11 to 13 are valid: e.g. 11*a* states that the statement 'A or not A' is always true, while 11*b* that 'both A and not A' is never true.

Let us look at the De Morgan laws. $\overline{A \vee B}$ means 'not either A or B', which is of course equivalent to 'not A and not B', which is $\overline{A} \wedge \overline{B}$. Similarly $\overline{A \wedge B}$ means 'not both A and B' and is equivalent to 'either not A or not B', which is $\overline{A} \vee \overline{B}$.

We have seen that the algebra of statements we have given obeys exactly the same laws as the algebra of sets. This is no accident: the logical reasoning involved is the same in both cases, and this type of algebra expresses basic logical ideas. A system obeying our laws is called a 'Boolean algebra' in the strict sense: the study of Boolean algebras is fruitful and has important applications. We have here an important example of the unifying aspect of mathematics: two subjects, set theory and the logic of statements, which appear to have little connection, are shown to be much the same, and may both be reduced to the same type of algebra, a process which retains the logically important aspects while discarding extraneous things such as the individual words of the statements.

As an example of the use of the algebra of statements, let us interpret the three results of §3.2 in this way. These have been proved for sets but, as the algebras are identical, they must be valid also for statements.

(*a*) $A \vee \overline{B}$ means: 'either A is true or B is false' while $\overline{B-A}$ means (interpreting '$-$' in the obvious sense): 'it is not true that B is true and A is false', which is clearly equivalent to the other.

(If we think that this result is more obvious than the other we could start from the statement form and deduce the set form.)

(*b*) $A \subseteq \overline{B} \wedge \overline{C}$ means that: 'the truth of A implies that neither B nor C is true', and our result tells us that this whole statement implies that not all of A, B and C can be true. But the converse does not logically follow.

(*c*) $\overline{\overline{A} \wedge \overline{B}}$ means: 'A and B are not both false', and this is equivalent to $A \vee B$. that is to: 'A or B is true'.

3.7. Syllogisms

In a syllogism we are given two or more statements as premises, from which we are able to deduce logically one or more further statements. A simple example is given below.

Example 1. We are given that:

(*a*) every monkey likes bananas;

(*b*) animals that like bananas have good teeth.

From this we can deduce that every monkey has good teeth.

The logical process in this example is easy to follow, but in more complicated cases it can be difficult or impossible to discover the conclusion without some symbolical aid. Set theory provides such an aid. The theory used is very easy, and we will see how it works in example 2, which is taken from Lewis Carroll's *Symbolic Logic.* Note that in this section capital letters again denote sets (for which certain statements are true), as distinct from in §3.6, where they stood for the statements themselves.

Example 2. Given the following premises find the conclusions:

(*a*) babies are illogical;
(*b*) nobody is despised who can manage a crocodile;
(*c*) illogical persons are despised.

We could solve this by ordinary reasoning, but a set theoretic method is easier and surer. We let A be the set of all persons who can manage a crocodile, B be the set of all babies, C the set of those people who are despised and D the set of logical people. (Note that if we like we may take the universal set to be the set of all persons):

(*a*) may be expressed as $B \subseteq \bar{D}$;
(*b*) gives us that $A \cap C = \varnothing$;
(*c*) gives us that $\bar{D} \subseteq C$.

In a Venn diagram these become as shown, from which we deduce that $A \cap B = \varnothing$, i.e. that babies cannot manage crocodiles, and that $A \cap \bar{D} = \varnothing$, i.e. $A \subseteq D$, or that all persons who can manage crocodiles are logical.

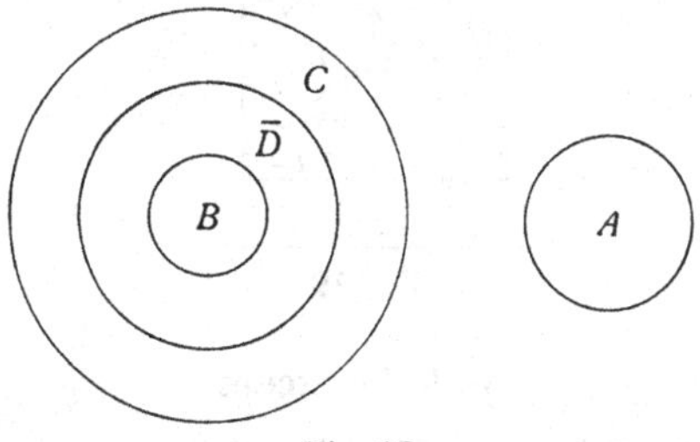

Fig. 17

All syllogisms may be reduced to this type of set theory. We may also by this method detect inconsistencies in the statements of a person, as in example 3.

Example 3. Show that the following statements are inconsistent:

(*a*) no unripe fruit is good for you;
(*b*) all these apples are good for you;
(*c*) no fruit, grown in the shade, is ripe;
(*d*) these apples were not grown in the sun.

Let: A be the set of all fruit grown in the shade,
B be the set of all ripe fruit,
C be the set of these apples,
D be the set of fruit that is good for you.

The statements become:

(*a*) $\bar{B} \cap D = \varnothing$, i.e. $D \subseteq B$;
(*b*) $C \subseteq D$;
(*c*) $A \cap B = \varnothing$;
(*d*) $C \subseteq A$.

We deduce that $C \subseteq B$ and also have $C \subseteq A$. Hence $A \cap B \supseteq C$ which contradicts (*c*). (We assume $C \neq \varnothing$.)

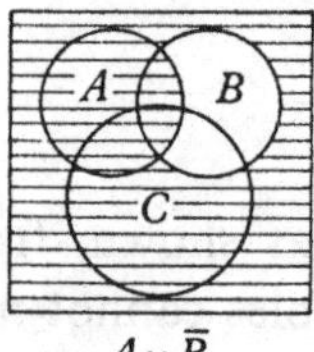

$A \cup \bar{B}$

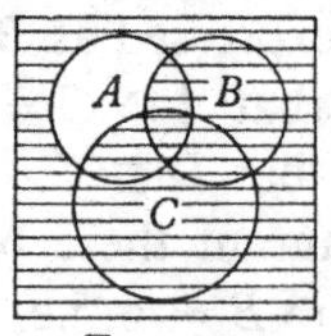

$\bar{A} \cup B \cup C$

Fig. 18

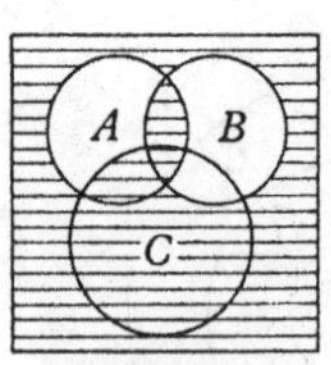

Fig. 19

Worked exercises

Since several examples have been worked in the text of this chapter we content ourselves with giving two more only.

1. Find the complement of $(A \cup \bar{B}) \cap (\bar{A} \cup B \cup C)$ and write down the dual result.

$A \cup \bar{B}$ and $\bar{A} \cup B \cup C$ are shown in the Venn diagrams of figure 18, from which we see that $(A \cup \bar{B}) \cap (\bar{A} \cup B \cup C)$ is as shown in figure 19.

Hence its complement is $(\bar{A} \cap B) \cup (A \cap \bar{B} \cap \bar{C})$. (This could also be written down using the law 14.) The dual result is that

$$\overline{(A \cap \bar{B}) \cup (\bar{A} \cap B \cap C)} = (\bar{A} \cup B) \cap (A \cup \bar{B} \cup \bar{C}).$$

Note that this is also the complementary result.

2. I have some bricks, each painted with one or more of the 3 colours, red, yellow and blue. 17 contain some red, 23 some yellow and 13 some blue. 6 contain both red and yellow, 9 yellow and blue and 4 blue and red, while just 2 have all three colours included. How many bricks have I?

Suppose the set of all those containing some red is R, the set of those with yellow is Y and of those with blue B. Let $N(X)$ be the number of elements in the set X. Then $N(R \cup Y) = N(R)+N(Y)-N(R \cap Y)$, by considering the Venn diagram.

Hence

$$\begin{aligned} N(R \cup Y \cup B) &= N(R \cup Y)+N(B)-N((R \cup Y) \cap B) \\ &= N(R)+N(Y)-N(R \cap Y)+N(B) \\ &\quad -N((R \cap B) \cup (Y \cap B)) \\ &= N(R)+N(Y)-N(R \cap Y)+N(B) \\ &\quad -N(R \cap B)-N(Y \cap B)+N(R \cap B \cap Y \cap B) \\ &= \Sigma N(R)-\Sigma N(Y \cap B)+N(R \cap Y \cap B). \end{aligned}$$

Putting in our values for $N(R)$ etc. we have that the total number of bricks $= N(R \cup Y \cup B) = 17+23+13-6-9-4+2 = 36$.

Exercise 3A

Prove **1–11**: (*a*) by use of Venn diagrams, and (*b*) by consideration of elements.

1. $A \cap B \subseteq A \cup B$.

2. $A \cap B = A \cup B \Leftrightarrow A = B$.

3. $A \subseteq B \Leftrightarrow \bar{A} \supseteq \bar{B}$.

4. $A \subseteq B \Leftrightarrow A \cap \bar{B} = \varnothing$.

5. $(A-B)-C = (A-C)-(B-C)$.

6. $A \subset B$ and $B \subseteq C \Rightarrow C \not\subseteq A$. Is the converse implication true?

7. $(A \cap B) \cup C = A \cap (B \cup C) \Leftrightarrow C \subseteq A$.

8. $A \cap B \subseteq \bar{C}$ and $A \cup C \subseteq B \Rightarrow A \cap C = \varnothing$.

9. $A \subseteq \overline{B \cup C} \Rightarrow A \cap B = \varnothing$ and $A \cap C = \varnothing$.

10. $\overline{A \cap \bar{B}} \cup B = \bar{A} \cup B$.

11. $\overline{(A \cap X) \cup (B \cap \bar{X})} = (\bar{A} \cap X) \cup (\bar{B} \cap \bar{X})$.

12. Prove nos. **1**, **5** and **10** by the truth table method.

13. Prove nos. **10** and **11** by using the laws of §3.4.

14. Simplify the sets $A \cap \varnothing$, $A \cup \varnothing$, $A-\varnothing$, $A-A$, $\varnothing-A$.

15. Draw a Venn diagram showing the following categories of numbers: complex numbers, real numbers, negatives, integers, multiples of 3, multiples of 5, numbers of the form $a+bi$ with a and b integers.

16. Draw a Venn diagram showing the following groups of humans: Europeans, Welsh, white-skinned people, Russians, people who live on an island (rather than a main land-mass), black-skinned people.

17. Show by theorem 3.5.6 that in forming dual results by interchanging $\cap$ and $\cup$ we must replace $\subseteq$ by $\supseteq$.

18. Show by theorem 3.5.4(i) that the dual of $A-B$ is $\overline{B-A}$.

19. Write down the results dual to nos. **1–11.** Which of them are self-dual?

20. Prove laws $6a$, $8a$ and $8b$ of §3.4 by consideration of elements and by the truth table method.

21. Use laws $6(a)$ and $7(a)$ to prove that $A \cup (A \cap B) = A \cap (A \cup B)$.

22. Prove the De Morgan laws $14a$ and $14b$ by elements and by the truth table method.

23. Prove theorem 3.5.2 by another method.

24. Prove that if $A \cup B = B \; \forall A$, then $B = I$.

25. Prove that if $A \cap B = B \; \forall A$, then $B = \emptyset$.

26. Interpret the results in nos. **1–11** in terms of the algebra of statements.

Find the conclusions of the syllogisms in nos. **27–33.** (These are taken from Lewis Carroll's *Symbolic Logic.*)

27. (*a*) No potatoes of mine, that are new, have been boiled.
(*b*) All my potatoes in this dish are fit to eat.
(*c*) No unboiled potatoes of mine are fit to eat.

28. (*a*) All the old articles in this cupboard are cracked.
(*b*) No jug in this cupboard is new.
(*c*) Nothing in this cupboard, that is cracked, will hold water.

29. (*a*) Puppies that will not lie still are always grateful for the loan of a skipping rope.
(*b*) A lame puppy would not say 'thank you' if you offered to lend it a skipping rope.
(*c*) None but lame puppies ever care to do worsted-work.

30. (*a*) All members of the House of Commons have perfect self-command.
(*b*) No M.P., who wears a coronet, should ride in a donkey race.
(*c*) All members of the House of Lords wear coronets.

31. (*a*) No boys under 12 are admitted to this school as boarders.
(*b*) All the industrious boys have red hair.
(*c*) None of the day-boys learn Greek.
(*d*) None but those under 12 are idle.

32. (*a*) Things sold in the street are of no great value.
(*b*) Nothing but rubbish can be had for a song.
(*c*) Eggs of the Great Auk are very valuable.
(*d*) It is only what is sold in the streets that is really rubbish.

33. (*a*) No kitten that loves fish is unteachable.
(*b*) No kitten without a tail will play with a gorilla.
(*c*) Kittens with whiskers always love fish.
(*d*) No teachable kitten has green eyes.
(*e*) No kittens have tails unless they have whiskers.

Show that the sets of premises in nos. **34–36** are inconsistent:

34. (*a*) All white rabbits have long ears.
(*b*) No long-eared rabbits are called Timothy.
(*c*) I have a white rabbit called Timothy.

35. (*a*) None of the salmon in this river are more than two pounds in weight.
(*b*) This river contains all the salmon that I own.
(*c*) No salmon of less than two pounds is fit to eat.
(*d*) Not all my salmon are unfit to eat.

36. (*a*) All my uncles are policemen.
(*b*) Nobody who likes cats is to be despised.
(*c*) My uncle Samuel has red hair.
(*d*) No policeman dislikes cats.
(*e*) A man who is not despised cannot have red hair.

Exercises 3B

1. Prove that $(A \cap B \cap C) \cup (\bar{A} \cap B \cap C) \cup \bar{B} \cup \bar{C} = I$, the universal set.

2. Prove that

$$(A \cap X) \cup (B \cap \bar{X}) \cup (C \cap X) \cup (D \cap \bar{X}) = [(A \cup C) \cap X] \cup [(B \cup D) \cap \bar{X}].$$

3. Prove that it is possible to find a set X such that $X \cup A = B$ if and only if $(A \cup B) - (A \cap B) \subseteq B$.

4. Explain the reason why $\subseteq$ is not an equivalence relation between pairs of sets.

5. In the set of positive integers $x|y$ means 'x is a factor of y'. Show that $|$ satisfies the same 3 laws as inclusion, given in §3.4, laws 1–3.

(A relation which satisfies these is known as a *partial ordering* of the elements.)

6. Define $A+B = (A \cup B) - (A \cap B)$. Prove that:

(i) $(A+B)+C = A+(B+C)$;
(ii) $A+\varnothing = A$;
(iii) $A+A = \varnothing$.

7. Deduce theorem 3.5.4 for the case $A = I$, the universal set, from laws 11–14.

8. Replacing '$-$' according to the law that $A-B = A \cap \bar{B}$ prove theorem 3.5.4(ii)–(vi) from laws of §3.4.

9. Label each of the 8 regions in figure 6 on p. 28 in terms of $A, B, C, \bar{A}, \bar{B}, \bar{C}$ and $\cap$ only (i.e. without using $\cup$ or $-$).

10. Prove laws 1–3 by theorem 3.5.6.

11. Draw a diagram similar to figure 6 for 4 sets A, B, C, D and label all 16 areas. (Do not use circles to represent the sets.)

12. Find simpler expressions for the complements of the following and write down the dual results:

(i) $(A \cup B \cup \bar{C}) \cap (A \cup \bar{B})$;
(ii) $(\bar{A} \cap B) \cup (A \cap \bar{B})$;
(iii) $A \cup (B \cap (C \cup D))$.

13. Prove that $(A \cap B) \cup (A \cap \bar{B}) \cup (\bar{A} \cap B) \cup (\bar{A} \cap \bar{B}) = I$ by the Distributive Law and explain it in terms of a Venn diagram.

14. Prove that $B = (A \cap \bar{B}) \cup (\bar{A} \cap B) \Leftrightarrow A = \varnothing$. (Poretsky's Law.)

15. In a certain village of 1000 houses, 750 have a car, 800 have a refrigerator, 850 have a telephone and 950 have a radio. What is the least number of houses that have all four?

16. Of 90 rabbits it is found that 57 are piebald, 43 are male and 32 have long ears. 19 males have long ears, 21 piebald have long ears and there are 30 male piebalds. 10 of the rabbits are non-piebald, short-eared females. How many male long-eared piebalds are there?

4

THE SET OF INTEGERS

4.1. The natural numbers

The positive integers, or 'natural numbers', are the simplest elements with which we deal in mathematics. They are the first abstraction that we make from our experience when we first, as young children, begin to learn the rudiments of arithmetic. They also come first historically—the very beginning of arithmetic and of mathematics is counting, although in some primitive peoples this proceeds only a small way: for example, as one, two, then many.

The idea of a whole number is comparatively simple, at any rate as compared with other mathematical concepts. It required initially a great leap of abstraction to realise that what a pile of six apples and a group of six people have in common is the number 'six', especially since this was the first abstraction of this type to be considered, but the average child today does not experience very much difficulty in the idea, certainly less than he has in mastering the various extensions of the number system that we will deal with later. For a long time the natural numbers were the only type of number, indeed the only type of mathematical entity except possibly those involved in geometry, to be used: the isolation of the 'six' from the pile of six apples was the step that made arithmetic possible.

Although integers seem a fairly obvious notion, it is not easy to give a rigorous definition for them. Once the existence of the natural numbers is assumed then the definition of other types of numbers, fractions, negatives and even irrationals and complex numbers, follows more or less easily. The details of definition may be complicated, but the processes are mathematically satisfying and rigorous, and are above all mathematical processes. Any definition of integer, on the other hand, raises deep logical and philosophical problems, and in fact belongs more to philosophy than to mathematics. For this reason many mathematicians prefer to accept the integers as known and base their work on them: in Kronecker's words 'God made the integers; all the rest is the work of man'.

The modern investigation of the philosophical foundations of arithmetic started at the end of the last century, the leading investigators being Gottlob Frege (whose *Die Grundlagen der Arithmetik* was published in 1884), Bertrand Russell (Russell and A. N. Whitehead's *Principia Mathematica* was published in 1910), Georg Cantor and G. Peano. Probably the simplest set of postulates for the natural numbers are Peano's, which are roughly as follows:

(*a*) The natural numbers form a set such that every member has a 'successor', itself in the set, but there exists at least one member which is not itself a successor.

(*b*) If a subset of the natural numbers contains an element that is not itself a successor and also contains the successor of all elements of itself then it coincides with the *whole* set of natural numbers.

These imply that there is just one element which is not a successor. This is called '1', its successor is '2', the successor of 2 is '3' and so on.

The philosophical investigations of number need not detain us: we are primarily interested in the algebraic properties. The basic algebra of the natural numbers is:

(*a*) Any two natural numbers may be added, and the sum is itself a natural number, i.e. addition is possible within the set. Addition is Commutative and Associative. This means that we may add three (or more) natural numbers by adding the third to the sum of the first two.

(*b*) Any two natural numbers may be multiplied together to give a natural number. Product is Commutative and Associative.

(*c*) The Cancellation Laws hold, i.e. $x+y = x+z \Rightarrow y = z$, $xy = xz \Rightarrow y = z$.

(*d*) They are ordered, i.e. they may be arranged in an order $1 < 2 < 3 < 4 < \ldots$.

The natural numbers have the following *negative* properties:

(*a*) Subtraction *within the set* is not always possible. Thus 4–6 is meaningless in terms of positives, though $6-4 = 2$.

(*b*) Division is not always possible within the set.

(*c*) Equations, even linear equations, do not always have a solution, e.g. the equations $x+4 = 2$ and $3x = 2$ have no solutions within the set, though $x+2 = 4$ and $3x = 6$ do have.

Note that the *proof* of these properties from the postulates is quite long.

Because of the possibility of adding and multiplying we may use such shorthand notations as $2x$ for $x+x$, x^2 for $x.x$, and so on. The index laws, that $x^m.x^n = x^{m+n}$ and $(x^m)^n = x^{mn}$ are valid. Note that $2x$ may mean either $x+x$ or $2.x$, but these of course have the same value.

4.2. The complete set of integers

The natural numbers were at first and for a long time the only numbers used in arithmetic and algebra, but gradually the need was felt for extensions of the number system. Thus simple equations as in (*c*) above could not be solved, neither could the corresponding problems such as 'take away 4 apples from 2 apples' or 'divide 2 apples into 3 equal parts'. The first other numbers to be used were the fractions, probably because an intuitive meaning can be given to these as applied to apples, cakes and other common objects, while negative numbers were introduced much later. As late as the twelfth century we find the Arab mathematician Omar Khayyam rejecting negative roots as meaningless when solving cubic equations.

From our point of view it is useful to introduce the negative integers before the fractions. This involves using the idea of zero, itself an idea which took a long time to develop.

We may think of the negative integers in various ways. The easiest way of understanding them, and therefore the one usually used when teaching children about them for the first time, is by the geometrical idea of a 'directed number'. We represent the natural numbers by points in a line, spaced at equal intervals from an origin, and think of them in terms of line segments from the origin going to the right (say). A negative integer is then represented by a line segment measured in the opposite direction. Addition is performed by placing the segments end to end, and subtraction defined as addition of the negative. The zero is of course a zero line segment. We thus obtain a line of integers as in figure 20, which will be added to in the next chapter by the inclusion of fractions and irrational numbers.

We may also think of the negatives algebraically as follows. We think of the number 0 as being that number which when added to any integer does not alter it, i.e. $x+0 = x$ whatever the value of x. Then the negative integer $-x$ is that number which when added to x gives 0, i.e. $(-x)+x = 0$. Subtraction of a number is the addition of its negative, and we can now subtract any two integers one from the other.

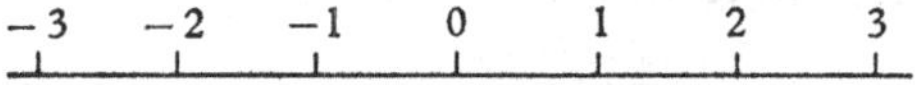

$-3 \quad -2 \quad -1 \quad 0 \quad 1 \quad 2 \quad 3$

Fig. 20

The ordinary rules for manipulating negatives follow easily from either of the above two methods of introducing them. Both methods have serious defects, however. The first is largely intuitive, using geometrical ideas, and it would be fairly difficult to make it completely rigorous, while the other assumes the existence of negatives and does not explain their formation. Once the idea of a negative is introduced and understood we seek a definition which is algebraic and abstract, so as not to depend on any properties of the natural numbers except the basic ones of addition and multiplication.

We now proceed to give a rigorous definition of integers (positive and negative) in terms of natural numbers. The procedure may seem needlessly complicated, but the precision required makes such a course necessary, and we will explain the steps carefully as we proceed. Throughout the work we bear in mind the properties we wish the finished product to have—it would be useless to define something which didn't behave as our intuitive ideas of negatives do.

Definition of integers

We consider all the ordered pairs (a, b) of natural numbers a and b. The statement that the pair is *ordered* means that it matters which of a and b we take first: (a, b) is not the same as (b, a).

We are going eventually to identify the pair (a, b) with the integer $a-b$. The set of ordered pairs will thus include all integers, positive negative or zero, but of course each will occur an infinite number of times. Thus we need to separate the pairs into classes, such that (a, b) and (c, d) will be in the same class if $a-b = c-d$. The best way to do this is by use of the equivalence classes theorem (theorem 2.4.1). Note that we will not use subtraction in the definition (since at present this is not always possible as we are dealing with natural numbers only).

Thus in the set S of ordered pairs (a, b) we form a relation R such that $(a, b)R(c, d)$ if $a+d = b+c$. We now need to show that R is an equivalence relation.

Reflexive. $(a, b)R(a, b)$ since we certainly have $a+b = b+a$.

Symmetric. If $(a, b)R(c, d)$ then $a+d = b+c$. Hence $c+b = d+a$ and so $(c, d)R(a, b)$.

Transitive. If $(a, b)R(c, d)$ and $(c, d)R(e, f)$ we have

$$a+d = b+c \quad \text{and} \quad c+f = d+e.$$

Hence $a+d+c+f = b+c+d+e$ and so, by the Cancellation Law of addition $a+f = b+e$ and $(a, b)R(e, f)$.

Thus R is an equivalence relation, and so S is divided into equivalence classes of mutually exclusive sets of pairs, with (a, b) and (c, d) in the same set if $a+d = b+c$. We now *define* an *integer* to be one of these equivalence classes.

Note 1. We have used the word *integer* here because we will eventually identify the classes with our usual positive negative or zero integers, and there is no point in coining a new word which would soon have to be discarded.

Note 2. It may seem strange to define a type of number as being a *class* of pairs. It is necessary to do so in order to use only the numbers with which we are already familiar, namely the natural numbers. We will soon be able to use the familiar notation for the new numbers and from thence all work using them will appear normal. The whole difficulty is to place negatives on a rigorous footing: once this is done their use is simple (as is the case with complex numbers).

We must now *define* addition and subtraction of our new integers, and we will discover that subtraction is always possible.

We define addition as follows: $(a, b)+(c, d) = (a+c, b+d)$ where (a, b) and (c, d) are any pairs from the classes which form the integers in question. To make the definition meaningful we need to show it to be independent of the particular pairs we choose. Let (a', b') and (c', d') be any two other pairs from the same classes. Then $a+b' = b+a'$ and $c+d' = d+c'$. Hence $a+c+b'+d' = b+d+a'+c'$ and so $(a+c, b+d), (a'+c', b'+d')$ are also in the same class. Hence given two integers, or classes, whichever pairs we choose when added according to our definition give rise to pairs which all are in the same class, and we may define their sum as being that class.

To define subtraction we first define the negative of any pair (a, b) as being (b, a) [intuitively $b-a = -(a-b)$]. Then if (a, b) and (a', b') are in the same class, so that $a+b' = b+a'$, so are (b, a) and (b', a') and we have a good definition of the negative of a class. (We have not yet obtained the concept of a negative integer, but only that of the negative of a given integer, the integers themselves not yet being divided into positive and negative.) Subtraction of an integer is now defined to mean the addition of its negative: thus $(a, b)-(c, d)$ means $(a, b)+(d, c)$ and equals $(a+d, b+c)$, where all the pairs are representatives from the classes that define the integers in question. Subtraction is of course always defined whatever the integers are.

We now wish to identify our integers as defined above with the intuitive idea of positive and negative integers. We divide the pairs of our set S into three subsets according as $a > b$, $a = b$ or $a < b$. We now show that all the pairs in an equivalence class belong to the same subset. For if $(a, b)R(c, d)$ then $a+d = b+c$ and so $d <, =, > c$ according as a $>, =, < b$, i.e. $c >, =, < d$ as a $>, =, < b$. Thus we have three types of integers, according as $a >, =, < b$.

$a > b$. We will identify these with the positive integers. To do this we must use the idea of subtraction of a smaller natural number from a larger. This does not involve a circular argument, since subtraction *in this case* does not depend on the idea of a negative and may be defined solely in terms of the natural numbers, depending on which philosophical basis we chose for these. (For example, in terms of Peano's postulates, if $x > 1$,

$x-1$ is the number of which x is the successor $x-2 = (x-1)-1$, and so on.) Then if $a > b$, $a-b$ is meaningful and we identify (a, b) with the natural number $a-b$ and say that (a, b) is a *positive* integer. All pairs in the same class give rise to the same natural number, since if $a+d = b+c$ then $a-b = c-d$. The sum of two pairs corresponds to the sum of the corresponding natural numbers, and so no confusion will arise by this identification. We see in fact that the set of all integers (defined as classes of pairs) such that $a > b$ behaves exactly the same as the set of natural numbers which we started with: for all practical purposes they are the same and we may drop our pair notation and now speak of the *positive integer* x (which stands for the class of all pairs (a, b) such that $a-b = x$).

$a < b$. If $a < b$ then (a, b) is the negative of the pair (b, a) which is a pair corresponding to a positive integer. If (b, a) corresponds to the positive integer x, we write its negative as $-x$, so that (a, b) corresponds to $-x$. Thus we have now obtained a rigorous definition of negative integers as being the negatives (in the precise meaning of this term that we gave) of positive integers (which we were familiar with at the start of the present piece of work (calling them natural numbers to avoid confusion) and later identified with certain classes of pairs).

$a = b$. All pairs (a, a) belong to the same class ($a+b = a+b$ and so $(a, a)R(b,b)$). We call this class the zero integer, or briefly the *zero*, and denote it by 0. We can easily verify that $0+x = x$ for any integer x, and $x+(-x) = 0$.

We have now obtained a complete set of positive, zero and negative integers, and the usual rules of combination, such as the Commutative Law that $x+y = y+x$, the Associative Law that $(x+y)+z = x+(y+z)$, and rules for the negative such as $-(-x) = x$, can easily be shown to apply from the number pairs definition above. It only remains to define multiplication (division is of course not always possible). This could be done in terms of the number pairs, but it is simpler to use the multiplication of positive integers, which is already known, and to define $x.(-y)$ to be $-xy$ (where x and y are positive), and $(-x).(-y)$ to be xy. $0.x$ is of course 0. With these definitions all the rules for multiplication may easily be shown to hold, and we have obtained our set of integers and all their properties.

The above treatment shows the care that must be taken in defining new algebraic structures (i.e. sets in which we can perform some or all of the rules of combination) in a completely rigorous manner in terms of known structures. We started with the positive integers and have produced a new system, more complete than the first since we can now *always* subtract, purely from the positive integers and their rules of combination. This type of process is a common one in abstract algebra and will repay very careful study, though it should not of course be committed to memory save in broad outline. Notice that the process is mathematical, unlike any definition of positive integers, which inevitably becomes more philosophical. An even more important example of the process, the formation of rational numbers from the integers, is given a thorough treatment in §5.1. The basic algebra of the complete set of integers is then:

Addition and subtraction. These are always possible within the set. The Cancellation Law of addition holds, and any equation of the form $x+a = b$ has a unique solution within the set.

Multiplication. This is always possible within the set, but division may be performed only in special cases. There is, however, a number 1 which behaves for multiplication rather like 0 does for addition, in that multiplication of any number by 1 leaves the number unchanged (such a number is called a *unity* or *neutral element* of the set). The Cancellation Law of multiplication is true, unless $x = 0$, and we also have the fact that there are no zero divisors, i.e. $xy = 0 \Rightarrow$ either $x = 0$ or $y = 0$. This latter property is important, being necessary, though not sufficient, for the uniqueness of factorisation of a positive integer into primes, as discussed in §4.6. A set in which we can add, subtract, and multiply (and in which the Commutative, Associative and Distributive Laws are true), which contains a unity and which has no zero divisors, is called an *integral domain.* The properties of such sets are studied in volume 2 and are somewhat similar to those of the integers. Note that the equation $ax = b$ does not always have a solution within the set, although it has sometimes, and there is never more than one solution.

Order. The integers are ordered:

$$\ldots < -2 < -1 < 0 < 1 < 2 < \ldots.$$

Thus for any two integers x and y it is always true that either $x < y$, $x = y$ or $x > y$. Notice that there is no least or greatest member of the set—the set is *unbounded.* When we represent the integers by points on a line $x < y$ means that x is to the left of y.

The relation satisfies the three laws:

(1) $x \leqslant x$;

(2) $x \leqslant y$ and $y \leqslant x \Rightarrow x = y$;

(3) $x \leqslant y$ and $y \leqslant z \Rightarrow x \leqslant z$.

(These are analogous to those given in §3.4 for inclusion of sets.)

4.3. Induction

From the ideas mentioned in §4.1 we see that every positive integer may be obtained by first taking the number 1, then choosing its successor 2 ($= 1+1$), taking the successor $2+1$ of 2 and so on. This property leads to the *principle of mathematical induction*, which states that if a certain property of the positive integers:

(*a*) can be proved to be true for the integer $k+1$, *provided* it is true for k, and

(*b*) is true for the integer 1,

then the property is true for all positive integers.

For by (*b*) it is true for 1 and so by (*a*) for 2, hence using (*a*) again it is true for 3, and so on. But this process finally includes any integer we care to name, and so the property is true for any integer, and so for all integers, where by 'integer' here we mean positive integer.

Example 1. Prove by induction that $\sum\limits_{r=1}^{n} r = \frac{1}{2}n(n+1)$.

(*a*) Assume that $\sum\limits_{1}^{k} r = \frac{1}{2}k(k+1)$. Then

$$\sum_{1}^{k+1} r = \tfrac{1}{2}k(k+1)+(k+1) = (k+1)(\tfrac{1}{2}k+1)$$
$$= \tfrac{1}{2}(k+1)(k+2);$$

(*b*) $\sum\limits_{r=1}^{1} r = \frac{1}{2}.1.2 = \frac{1}{2}.1(1+1)$.

Hence the result is true for $n = 1$ and so for $n = 2, 3, \ldots$, and thus is true for all positive integers n.

Induction is an extremely powerful method. In order to use it we must either guess or know the result to be proved, but this is often fairly simple, and then the application in (*a*) and (*b*) is usually straightforward.

Some modifications of the basic principle are shown below.

Example 2. (Induction on a property of the entities for which the theorem is proved, the entities not being themselves integers.)

Assuming that every polynomial equation with complex coefficients has at least one real or complex root, to prove that a polynomial of degree n has exactly n roots, possibly repeated.

(*a*) Assume the theorem true for all polynomials degree k. If $P(x)$ is any polynomial degree $k+1$, $P(x) = 0$ has at least one root α by hypothesis, and so $P(x) \equiv (x-\alpha)Q(x)$, where $Q(x)$ is a polynomial degree k, obtained by dividing out. But by the inductive hypothesis $Q(x)$ has exactly k roots, and so $P(x)$ has $k+1$.

(*b*) Any polynomial equation degree 1 is of the form $ax+b = 0$, and has precisely one root $-b/a$.

The theorem follows by induction.

Example 3. (Assuming the result true for *all* positive integers $\leqslant k$.) Suppose $u_1 = 7$ and

$$u_n = u_{n-1}+2u_{n-2}+3u_{n-3}+ \ldots +(n-1)u_1 \quad \text{for all } n \geqslant 2.$$

Prove that u_n is divisible by 7.

(*a*) Assuming $u_k, u_{k-1}, \ldots, u_1$ all divisible by 7 it is immediate that u_{k+1} is divisible by 7.

(*b*) $u_1 = 7$ and is divisible by 7.

Hence u_2 is divisible by 7, and hence so is u_3 by (*a*), and so on.

Example 4. (Induction both ways so as to apply to all integers, positive and negative.)

If $u_{r+1} = 2u_r+1$ for all integral r, positive, negative or zero, prove that $u_n+1 = 2^{n-1}(u_1+1)$.

(*a*) Assume the result true for k, i.e. $u_k+1 = 2^{k-1}(u_1+1)$. Then

$$u_{k+1}+1 = 2u_k+2 = 2(u_k+1) = 2^k(u_1+1).$$

Also $u_{k-1}+1 = \frac{1}{2}(u_k-1)+1 = \frac{1}{2}(u_k+1) = 2^{k-2}(u_1+1)$.

Hence the result is true for both $k+1$ and $k-1$.

(*b*) This is trivial, for $n=1$.

It follows from (*a*) that the result is true for $n = 2$ and 0, and so for 3 and -1, and so on, for all positive and negative integers.

Example 5. (To prove a result false by *reductio ad absurdum*.)

Prove that $3^{2n}+5$ is never divisible by 8.

Suppose that $3^{2k}+5$ *is* divisible by 8.

If $u_n = 3^{2n}+5$ we have that $u_k-5 = 9(u_{k-1}-5)$, and so $u_k-u_{k-1} = 8u_{k-1}-40$ and is divisible by 8. Hence if u_k is divisible by 8, so is u_{k-1}, and thus the hypothesis that u_k is implies that u_{k-1} is, and so that u_{k-2} is divisible by 8, and so on. Hence it implies that u_1 is divisible by 8. But $u_1 = 14$ and we have a contradiction.

4.4. Divisibility

We have said that it is always possible to add, subtract or multiply two integers and obtain a third, but that division is not always possible. In certain cases, however, we can divide the integer a by b and obtain an integer c. Then $a = bc$. We say that a is *divisible* by b, a is a *multiple* of b, and b is a *factor* or *divisor* of a. We denote this property by writing b/a. (It is understood that neither a nor b is zero.)

Divisors of 1 are called *units* (distinguish carefully between these and the *unity* 1). The only units are of course 1 itself and -1.

If b is a factor of a then so is $-b$, and b, $-b$ are also factors of $-a$. Thus the property of divisibility is unaltered when we multiply either of the numbers concerned by a unit.

If the only factors of p are ± 1, $\pm p$, and $|p| > 1$, then p is called a *prime*. The study of primes is of the highest importance in the theory of numbers. An integer that is not prime is called *composite*.

Theorem 4.4.1. *b/a and $a/b \Rightarrow a = \pm b$.*

Since b/a we have $a = bc$ for some c. Since a/b we have

$b = ad$ for some d. Hence $a = acd$ and so, by the cancellation law, since $a \neq 0$, $cd = 1$. Hence c is a unit and so is ± 1, giving $a = \pm b$.

4.5. Highest common factor

The reader will be familiar with the idea of the highest common factor (H.C.F.) or greatest common divisor of two positive integers. In elementary arithmetic this is usually defined as the greatest integer which is a factor of both, and is found by expressing the given integers in terms of their prime factors and selecting the necessary common factors. We will introduce it by a rather different method which may seem unnecessarily obscure, but which is important theoretically and which will carry over to sets other than the integers, particularly to polynomials, as in chapter 7. Among other reasons for not using the elementary method of expressing the integers as products of their prime factors is that it has not been proved, and is by no means obvious, that such an expression exists and is unique—this latter fact will be proved in the next section, and the properties of the H.C.F. will be used there in the proof. Thus our method avoids a circular argument.

Definition. *The integer h is an* H.C.F. *of the two integers a and b (positive, negative but not zero) if*

(i) $h|a$ *and* $h|b$;

(ii) *if* $c|a$ *and* $c|b$ *then* $c|h$.

This definition says nothing about the existence or uniqueness of an H.C.F. The existence will be shown later in the section, when a method is given for finding one, while uniqueness in the strict sense does not quite hold, for if h is an H.C.F. then so is $-h$. Within this limitation the H.C.F. must be unique as is shown below.

Theorem 4.5.1. *If a and b have two* H.C.F.'s, *h and h', then $h' = \pm h$.*

Both h and h' are factors of a and b and so by part (ii) of the definition we have $h'|h$ and $h|h'$. The result follows by theorem 4.4.1.

Note that the definition does not restrict any of our integers to being positive. There are two H.C.F.'s—one positive and the other negative, and it is customary to call the positive one *the* H.C.F. Although this will in fact be the greatest of all common factors of a and b, this is not its fundamental property: the important point is that it is a multiple of *any* common factor of a and b (by (ii)).

The positive H.C.F. of a and b is denoted often by (a, b). If $(a, b) = 1$ the integers a and b are called *co-prime*.

We will now show how to find an H.C.F. by the method known as Euclid's algorithm. ('Algorithm' means 'method of computation' and is derived from the name of the Arab algebraist Al-Khowarizmi.) In practice this is only used for a pair of large and difficult numbers, but it is of great theoretical importance and is commonly used to find the H.C.F. of two polynomials, where factorisation may be a difficult or impossible process. The basis of the method is the so-called 'division algorithm'.

Theorem 4.5.2. *The division algorithm.*

Given any two integers a *and* b, *with* $b > 0$, *there exist unique integers* q *and* r, *with* $0 \leqslant r < b$, *such that* $a = bq+r$.

This states formally the familiar process of division to obtain a quotient (q) and remainder (r). It will probably seem obvious, and the proof which follows may well be omitted if found difficult.

There exists at least one multiple m of b such that $a-bm \geqslant 0$. For if $a \geqslant 0$ then $m = 0$ gives such a multiple, while if $a < 0$, $m = -|a|$ does so as $b \geqslant 1$.

Now consider the set of all values of $(a-bn)$ which are $\geqslant 0$. By the above this set is not empty. Hence it must possess a least member (this is a property of the positive or zero integers). Let this least member be r and arise when $n = q$. Then $a = bq+r$, and $r \geqslant 0$. It remains to show that $r < b$. If $r \geqslant b$ we have $r-b \geqslant 0$. But $r-b = a-b(q+1)$ and so is a member of our set which is less than r, which is impossible. Hence $r < b$.

q and r are unique. For suppose q' and r' were another pair of integers with the same property. Then $a = bq+r = bq'+r'$. Hence $r-r' = b(q-q')$, and so $r-r'$ is a multiple of b. But $0 \leqslant r < b$ and $0 \leqslant r' < b$ and so $|r-r'| < b$, and so must be zero. Hence $r = r'$ and so $q = q'$.

Euclid's algorithm. *To find an* H.C.F. *of a and b.*

Since it is clear from the definition that $(a, b) = (a, -b)$ we may suppose $b > 0$ and so may apply the division algorithm, obtaining

$$a = bq_1 + r_1, \quad \text{say, with } 0 \leqslant r_1 < b.$$

If $r_1 \neq 0$ apply the division algorithm to b and r_1, obtaining

$$b = r_1 q_2 + r_2 \quad \text{with} \quad 0 \leqslant r_2 < r_1.$$

If $r_2 \neq 0$, continue this process, obtaining a series of quotients and remainders. Since $b > r_1 > r_2 > r_3 > \ldots$ we must obtain a zero remainder, say r_{n+1}, after a finite number of steps. We then have the series of equations:

$$\begin{aligned} a &= bq_1 + r_1, \\ b &= r_1 q_2 + r_2, \\ r_1 &= r_2 q_3 + r_3, \\ &\ldots\ldots\ldots\ldots\ldots\ldots \\ r_{n-2} &= r_{n-1} q_n + r_n, \\ r_{n-1} &= r_n q_{n+1}. \end{aligned}$$

We will show that r_n is an H.C.F. of a and b.

(i) Working from the bottom upwards, we have first that r_n is a factor of r_{n-1}, and so by the penultimate equation it must be a factor also of r_{n-2}, and hence of r_{n-3} and so on. Hence it is a factor of b and so of a.

(ii) Suppose c/a and c/b. Then working downwards we have c/r_1, and hence c/r_2 and so on, obtaining finally that c/r_n.

Euclid's algorithm shows us that an H.C.F. exists for any two integers, and we have already shown it unique to within $\pm$. Thus the notation (a, b) and the expression '*the* H.C.F.' are justified.

An important deduction from Euclid's algorithm is the following theorem.

Theorem 4.5.3. *If $h = (a, b)$ then there exist integers s and t such that $h = sa + tb$.*

From the above

$$\begin{aligned} h = r_n &= r_{n-2} - r_{n-1}q_n \\ &= r_{n-2} - q_n(r_{n-3} - r_{n-2}q_{n-1}) \\ &= A_1 r_{n-3} + B_1 r_{n-2} \quad \text{where } A_1 \text{ and } B_1 \text{ are integers} \\ &= A_1 r_{n-3} + B_1(r_{n-4} - q_{n-2}r_{n-3}) \\ &= A_2 r_{n-4} + B_2 r_{n-3} \end{aligned}$$

and so on, obtaining finally

$$h = sa + tb.$$

Corollary. *If a and b are co-prime, so that $h = 1$, there exist s and t such that $1 = sa + tb$.*

Theorem 4.5.4. *The set $\{ma + nb\}$, where m and n range over the integers, consists of the multiples of h (each being repeated).*

Since h/a and h/b it also divides $ma + nb$. Thus the set includes only multiples of h. It includes *all* these multiples since the number $kh = (ks)a + (kt)b$ and is thus included.

The H.C.F. *of more than 2 integers*

A number h is an H.C.F. of the n integers $a_1, a_2, \ldots, a_n$ if

(i) h/a_i for all $1 \leqslant i \leqslant n$;

(ii) any factor of all the a_i's is a factor of h.

As in the case of 2 integers there cannot be more than one H.C.F., to within $\pm$. Also one certainly exists, as may be proved by induction as follows.

Suppose an H.C.F. g_k exists of $a_1, a_2, \ldots, a_k$, and consider $g_{k+1} = (g_k, a_{k+1})$. Then g_{k+1}/a_{k+1} and g_{k+1}/g_k. But g_k/a_i, $1 \leqslant i \leqslant k$, and so g_{k+1}/a_i also. Any factor of a_i, $1 \leqslant i \leqslant k+1$, is a factor of g_k by hypothesis, and also of a_{k+1}, and hence of g_{k+1}. Thus g_{k+1} is the H.C.F. of $a_1, a_2, \ldots, a_{k+1}$. Hence if an H.C.F. exists of the first k integers of the given set, one exists of the first $k+1$. But a_1 and a_2 have an H.C.F. Hence there is an H.C.F. of $a_1, a_2, \ldots, a_n$, by induction, and the above proof indicates one method of finding it.

4.6. Prime factorisation

In this section we prove the so-called 'fundamental theorem of arithmetic', that any integer can be expressed uniquely as the product of ± 1 and a set of positive primes. We first need the result of theorem 4.6.1, the proof of which itself depends on theorem 4.5.3.

Theorem 4.6.1. *If p is a prime and $p|ab$ then either $p|a$ or $p|b$.*

Suppose p is not a factor of a. Then p and a can have no common factors except ± 1, since p is prime. Hence $(p, a) = 1$ and so by theorem 4.5.3 there are integers s and t such that $1 = sp + ta$. Hence $b = sbp + tab$. But p is a factor of ab by hypothesis and so it is a factor of b.

Theorem 4.6.2. *The fundamental theorem of arithmetic.*

Any integer a may be expressed in the form $\pm p_1 p_2 p_3 \dots p_m$, where the p_i's are positive primes, and this expression is unique except for the order of the factors.

(*a*) *To prove possibility of prime factorisation*

We use induction, and suppose that prime factorisation is possible for all positive integers less than k. With this hypothesis we will prove it possible for k. Then since it is possible for 1 it is so for all positive integers by induction, and therefore for negative integers also (by changing the initial sign).

Suppose then that it is possible to factorise any integer less than k. Either k is a prime, in which case the factorisation is trivial, or it possesses a factor k_1 such that $k = k_1 k_2$ say, with k_1 and k_2 less than k. Then by hypothesis both k_1 and k_2 may be factorised, say $k_1 = \pm q_1 q_2 \dots q_s$ and $k_2 = \pm r_1 r_2 \dots r_t$. Thus $k = \pm q_1 q_2 \dots q_s r_1 r_2 \dots r_t$ and can therefore be factorised in the required way.

(*b*) *To prove uniqueness of prime factorisation*

Suppose we have two expressions for a. Let $a = \pm p_1 p_2 \dots p_m$ and $a = \pm q_1 q_2 \dots q_n$. Then both signs are the same since all the p's and q's are positive. p_1 is a factor of $q_1 q_2 \dots q_n$. By an immediate corollary of theorem 4.6.1 it follows that p_1 is a factor of some q_j, say q_α. But q_α is a prime. Hence $p_1 = q_\alpha$, since both are positive. Thus

$$\begin{aligned} p_1 p_2 \dots p_m &= q_\alpha q_1 \dots q_{\alpha-1} q_{\alpha+1} \dots q_n \\ &= p_1 q_1 \dots q_{\alpha-1} q_{\alpha+1} \dots q_n \end{aligned}$$

and, dividing by p_1 we have

$$p_2 \cdots p_m = q_1 \cdots q_{\alpha-1} q_{\alpha+1} \cdots q_n.$$

Repeating the process we find that $p_2 = q_\beta$ say and divide out. Continuing, we cannot use all the q's before we exhaust the p's and we cannot have any q's remaining when the p's are exhausted. Hence $m = n$, and each p is equal to some q. Thus the p's correspond to the q's in some order and so the prime factorisation is unique.

If we consider prime factorisation including the use of negative primes we see that it is unique within multiplication of any prime by ± 1, i.e. within multiplication by a unit. This is of course the most we can hope to achieve.

We are now in a position to find the H.C.F. by the elementary method. For two integers we express each in terms of its prime factors and merely take the product of all the common ones. Thus if $a = p_1 \ldots p_m q_1 \ldots q_n$ and $b = p_1 \ldots p_m r_1 \ldots r_l$ where no q is an r, then the H.C.F. is $p_1 \ldots p_m$, for this is a common factor and contains all common factors as divisors of it.

4.7. Lowest common multiple

Definition. *The integer l is an* L.C.M. *of the two integers a and b if*

(i) *$a|l$ and $b|l$,*

(ii) *if $a|c$ and $b|c$ then $l|c$.*

It is immediate as for the H.C.F. that there cannot be more than one L.C.M. (within $\pm$). To show that there *is* one we express a and b in terms of their prime factors. Then if a and b are as above (end of § 4.6) the L.C.M. is $p \ldots p_m q_1 \ldots q_n r_1 \ldots r_k$, since this obviously has the required properties.

The L.C.M. of n integers is defined analogously to the H.C.F.

Theorem 4.7.1. *If h and l are the* H.C.F. *and* L.C.M. *of a and b $hl = ab$.*

Let

$$a = p_1 \ldots p_m q_1 \ldots q_n, \quad b = p_1 \ldots p_m r_1 \ldots r_k. \text{ Then}$$

$$h = p_1 \ldots p_m \quad \text{and} \quad l = p_1 \ldots p_m q_1 \ldots q_n r_1 \ldots r_k$$

and the result follows.

Worked exercises

1. Find the H.C.F. h of $a = 1705$ and $b = 625$ by Euclid's algorithm and express h in the form $sa+tb$:

$$\begin{aligned}1705 &= 2\times 625+455,\\ 625 &= 1\times 455+170,\\ 455 &= 2\times 170+115,\\ 170 &= 1\times 115+55,\\ 115 &= 2\times 55+5,\\ 55 &= 11\times 5.\end{aligned}$$

Hence $h = 5$.

$$\begin{aligned}h = 5 &= 115-2\times 55\\ &= 115-2(170-1\times 115) = 3\times 115-2\times 170\\ &= 3(455-2\times 170)-2\times 170 = 3\times 455-8\times 170\\ &= 3\times 455-8(625-1\times 455) = 11\times 455-8\times 625\\ &= 11(1705-2\times 625)-8\times 625 = 11\times 1705-30\times 625.\end{aligned}$$

Hence $h = 11a-30b$.

2. If h is the least positive number of the set $\{sa+tb\}$ prove directly that h is the H.C.F. of a and b.

Any common factor of a and b is a factor of $sa+tb$ and so is a factor of h. To prove that h is itself a common factor, suppose that it is not a factor of a. Then by the division algorithm $a = qh+r$ where $0<r<h$. Then since $h = sa+tb$ we have $a = qsa+qtb+r$ and so $r = (1-qs)a-qtb$. Hence r can be expressed in the form $Sa+Tb$, and r is a positive number less than h, which is contrary to hypothesis. Hence h is a factor of a and similarly of b.

3. Let p be a prime >3. Assume the theorem that if $0<n<p$ then there are integers a and b such that $1 = an+bp$. Prove that if $1<n<p-1$ then $a \neq n$. Hence or otherwise prove that $(p-2)!-1$ is divisible by p.

(Cambridge Open Scholarship.)

(Since p is prime, if $1<n<p$, p and n are co-prime and the existence of a and b follows from theorem 4.5.3. If $n = 1$ then $a = 1$ and $b = 0$.)

If $1<n<p-1$ suppose $1 = an+bp$ with $a = n$. Then

$$-bp = n^2-1 = (n+1)(n-1).$$

Hence either $p/(n+1)$ or $p/(n-1)$ by theorem 4.6.1. But this is impossible since both $n+1$ and $n-1$ are less than p. Hence $a \neq n$.

Now let $n = p-2$. Then a and b exist with $1\leqslant a\leqslant p$, for we may always add or subtract multiples of p to a by adjusting b by the same multiple of $-n$. But a cannot equal 1 since then p would have to divide $n-1$. Also a cannot equal p, or $p-1$ (in the latter case p would have to divide $n+1$). Hence there is an $a = a_1$ say, with $1<a_1<p-1$.

By the first part $a_1 \neq p-2$. Let $a_1(p-2) = a_1 n = 1-b_1 p$. Now let $n = p-3$. As before $\exists\ a_2$, $1<a_2<p-1$ such that $a_2(p-3) = 1-b_2 p$,

and $a_2 \neq p-3$. But $a_2 \neq a_1$ since if so we would have, subtracting, $a_1(p-2-p+3) = (b_2-b_1)p$ which is impossible since $1<a_1<p$. Similarly $a_2 \neq p-2$.

We proceed to find a_3 such that $a_3(p-4) = 1-b_3p$ and as before $1<a_3<p-1$ and a_3 is not equal to a_2 or a_1, or to $p-2, p-3$ or $p-4$. We continue with $p-5$, etc., omitting a number if it has appeared already as an a_i, and finally we have the numbers 2, 3, ..., $p-2$ arranged in pairs (there is an even number of them since p is odd). Let the pairs be a_i and c_i say. Then $a_ic_i = 1-b_ip$. Hence

$$(p-2)! = \Pi a_ic_i = \Pi(1-b_ip)$$
$$= 1 + \text{a multiple of } p.$$

Hence $(p-2)!-1$ is divisible by p.

(The last part may be proved otherwise by Wilson's theorem (theorem 6.6.3). For we have $(p-1)!+1$ is divisible by p, i.e.

$$(p-1)(p-2)!+1 = p.(p-2)!-((p-2)!-1)$$

is divisible by p and the result follows.)

Exercises 4A

Prove 1–10 by induction.

1. $\sum_{r=1}^{n} r^2 = \frac{1}{6}n(n+1)(2n+1)$.

2. $\sum_{r=1}^{n} \frac{1}{r(r+1)} = 1-\frac{1}{n+1}$.

3. $\sum_{r=1}^{n} (r^2+1)r! = n(n+1)!$

4. $\sum_{r=1}^{n} (-1)^r \frac{r+1}{(2r+1)(2r+3)} = (-1)^n \frac{1}{4(2n+3)} - \frac{1}{12}$.

5. $5^{2n}-1$ is divisible by 24 for n a positive integer.

6. $\sum_{r=0}^{n-1} x^r = \frac{x^n-1}{x-1}$.

7. $\sum_{r=1}^{n} rx^{r-1} = \frac{1-(n+1)x^n+nx^{n+1}}{(x-1)^2}$.

8. If $u_n = 3u_{n-1}+18u_{n-2}$ with $u_1 = 6$, $u_2 = 9$, prove that u_n is divisible by 3^n.

9. Prove the binomial theorem that $(a+x)^n = \sum_{r=0}^{n} {}_nC_r\, a^{n-r} x^r$. (Assume that ${}_nC_r + {}_nC_{r-1} = {}_{n+1}C_r$.)

10. Prove Leibniz's theorem that

$$\frac{d^n}{dx^n}(uv) = \sum_{r=0}^{n} {}_nC_r \frac{d^r u}{dx^r}\frac{d^{n-r} v}{dx^{n-r}}.$$

11. Prove that there is no greatest prime by considering $(n!+1)$ if n *were* the greatest prime.

12. Prove that if n is not prime then it has at least one prime factor $\leqslant \sqrt{n}$.

13. Show that the numbers from $(n+1)!+2$ to $(n+1)!+(n+1)$ give a series of n consecutive composite numbers.

14. Find the H.C.F. of the pairs in **14–19** by Euclid's algorithm and express it in the form $(sa+tb)$:

14. $a = 1625$, $b = 858$.

15. $a = 2904$, $b = 12{,}000$.

16. $a = 10{,}759$, $b = 2401$.

17. $a = 625$, $b = 1639$.

18. $a = 5307$, $b = 9150$.

19. $a = 135$, $b = 686$.

Find the H.C.F. and L.C.M. of the numbers in **20–22** by expressing them in terms of their prime factors and express the H.C.F. in the form

$$(sa+tb+uc+\ldots).$$

20. 16, 100, 350.

21. 78, 364, 715.

22. 105, 343, 238, 259.

23. Of which integers is 0 a divisor? Which integers are divisors of 0? Why do we except 0 in our definition of H.C.F.?

24. Prove that $(a, b) = (a, -b) = (-a, b) = (-a, -b)$.

25. Prove that if $(a, c) = 1$ and c/ab then c/b.

26. Prove that if $(a, c) = 1$ and a/m and c/m then ac/m.

27. Prove that if a/b and a/c then $a/(b+c)$.

28. Prove from the definition that $(ab, ac) = a.(b, c)$.

29. If $1 = sa+tb$ for some integers s and t show that $(a, b) = 1$.

30. Prove that if $(a, c) = (b, c) = 1$, $(ab, c) = 1$.

Exercises 4B

1. Assuming $\sum_{r=1}^{n} r = \frac{1}{2}n(n+1)$ prove by induction that $\left(\sum_{r=1}^{n} r\right)^2 = \sum_{r=1}^{n} r^3$ and deduce that $\sum_{r=1}^{n} r^3 = \frac{1}{4}n^2(n+1)^2$.

2. Show that a plane is divided by n straight lines, of which no two are parallel and no three meet in a point, into $\frac{1}{2}(n^2+n+2)$ regions. Consider the same problem with the plane replaced by the surface of a sphere and the lines by great circles of which no three meet in a point.

Into how many regions is space divided by n planes, of which no two are parallel, no three meet in a line and no four meet in a point?

(Cambridge Open Scholarship.)

3. Prove that
$$(1-\tfrac{1}{4})(1-\tfrac{1}{9})\ldots\left(1-\frac{1}{n^2}\right) = \frac{n+1}{2n} \quad (n \geqslant 2).$$

4. A finite sequence of real numbers $u_0, u_1, \ldots, u_n$ satisfies
$$(u_{k+1}-2u_k)^2 = 1, 0 \leqslant k < n.$$
Show that $u_n - 2^n u_0 + 2^n$ is a positive integer.

What values may this integer take? (Cambridge Open Scholarship.)

5. The functions $f_n(x)$ are defined thus:
$$f_0(x) = 1, \quad f_n(x) = (-\tfrac{1}{2})^n e^{x^2} \frac{d^n}{dx^n}(e^{-x^2}) \quad (n \geqslant 1)$$

Show that $f_n(x) = xf_{n-1}(x) - \tfrac{1}{2}f'_{n-1}(x)$ if $n \geqslant 1$, and deduce that $f_n(x)$ is a polynomial of degree n with leading coefficient 1.

(Cambridge Open Scholarship.)

6. A sequence $u_0, u_1, \ldots$ is defined by $u_0 = 3$, $u_{n+1} = (2u_n+4)/u_n$. Prove that

(i) $3 \leqslant u_n \leqslant \frac{10}{3}$ for all n;

(ii) $|u_n - a| \leqslant (\frac{4}{9})^n(a-3)$ for all n, where a is the positive root of $x^2-2x-4 = 0$. (Cambridge Open Scholarship.)

7. Prove that if h is the H.C.F. of a, b, c, it can be expressed in the form $h = sa+tb+uc$ for suitable s, t, u. Prove the analogous result for the H.C.F. of n integers.

8. Write out a rigorous proof of theorem 4.5.3 (by the method of the text) using induction.

9. Prove that if h, l are the H.C.F. and L.C.M. of a, b, c,
$$\begin{array}{ll} h_1, l_1 & \text{of } b, c, \\ h_2, l_2 & \text{of } c, a, \\ h_3, l_3 & \text{of } a, b, \end{array}$$
then $l^2h_1h_2h_3 = h^2l_1l_2l_3 = hlabc$.

10. If ab is a square and if $(a, b) = 1$, with a, b both positive, show that both a and b are squares.

11. If p and q are primes and $p \neq q$, show that the necessary and sufficient condition for us to be able to divide a circle into pq equal parts using ruler and compasses only is that we can do so for both p and q equal parts. (Use theorem 4.5.3.)

[This gives the condition that we can draw a regular polygon of pq sides by ruler and compasses only. It can be shown that the only primes or powers of primes for which this is possible are 2^n or $2^{2^n}+1$ provided the latter is prime, and hence the only polygons are those with sides 2^n, $2^{2^n}+1$ if prime, or products of these. $2^{2^n}+1$ is prime if $n = 0, 1, 2, 3, 4$—giving 3, 5, 17, 257, 65537.]

12. $s(n)$ denotes the sum of all the positive factors of n. Prove that, if m and n are co-prime, then $s(mn) = s(m).s(n)$.

5

OTHER SETS OF NUMBERS

5.1. The rationals

We have seen that although we can add, subtract or multiply two integers to obtain another integer the same is not true for division, and hence that such simple equations as $ax = b$ are not always solvable in terms of integers. In concrete terms there is no solution in whole cakes of the problem: divide 3 cakes equally among 5 people. Because of this it was found necessary to introduce fractions or, as we will call them, rational numbers. This step was taken very early in the development of arithmetic, far earlier than that of introducing negatives. The Greeks had a highly developed theory of ratios, although they tended to think of them in geometrical terms, and as early a mathematical people as the Babylonians sometimes used fractions.

We may think of the rational numbers in several ways. The intuitive method is to consider a fraction such as, for instance, $\frac{3}{5}$ to be obtained by dividing a cake into 5 equal parts and taking 3 of these. It is easily seen to be the result of dividing 3 cakes into 5 equal parts. Negative fractions follow as for negative integers.

We may also think of fractions as represented on our line of integers. The fraction $\frac{22}{7}$, as an example, would be written as $3\frac{1}{7}$ and obtained by dividing the segment from 3 to 4 into 7 equal parts and taking the point at the end of the first part.

We may define rationals in an analogous way to the algebraic method of defining the negatives. The number $1/n$, often called the *inverse* of n, is that number which when multiplied by n gives the unity 1. Then m/n means $1/n$ multiplied by m, where of course m and n are any integers, $n \neq 0$. (n cannot be zero since no number multiplied by 0 can give 1 and so $1/0$ is meaningless.) Notice here the analogy between the roles of 0 and $-x$, and of 1 and $1/x$ in the processes of addition and multiplication, respectively.

The ordinary notation and rules for manipulating fractions follow easily from any of the above methods of introducing them. All the methods given, however, suffer from the defect of not being completely algebraic or, in the case of the last, from not explaining the formation of the rationals. Once the idea of a rational number is introduced and understood we seek a method of definition which is algebraic and abstract, so as not to depend on any properties of integers except the basic ones of addition and multiplication.

We now proceed to give a rigorous definition of rationals in terms of integers. The advantage of this type of approach lies in the fact that it can be applied to sets other than the integers: later in this volume we will use it for polynomials, while in volume 2 it will be shown to apply even more generally.

Definition of the rationals

We start by considering the set of all ordered pairs of integers (a, b) with the proviso that $b \neq 0$. The fact that the pair is *ordered* means that it matters which integer we take first, so that (a, b) is not the same as (b, a). The pair (a, b) will of course be identified later with the rational number a/b, and we must have the condition that $b \neq 0$ since a fraction with zero denominator is meaningless.

We now wish to express the fact that $a/b = \lambda a/\lambda b$, i.e. that $a/b = c/d$ if and only if $ad = bc$. We do this by setting up an equivalence relation among the pairs of our set such that all equal factions will lie in the same equivalence class. We will then define a rational number as one of our equivalence classes.

Thus we set up a relation such that $(a, b)R(c, d)$ if and only if $ad = bc$. We must show that this is an equivalence relation.

Reflexive. $(a, b)R(a, b)$ since we certainly have $ab = ba$.

Symmetric. If $(a, b)R(c, d)$ then $ad = bc$ and so $cb = da$, hence $(c, d)R(a, b)$.

Transitive. If $(a, b)R(c, d)$ and $(c, d)R(e, f)$ we have that $ad = bc$ and $cf = de$. Hence $acdf = bcde$. But $d \neq 0$ and so either $c = 0$ or $af = be$. But if $c = 0$ we must have $ad = de = 0$ and $d \neq 0$. Thus $a = e = 0$ and we have in all cases that $af = be$. Hence $(a, b)R(e, f)$.

Our relation is therefore an equivalence relation and so it divides the pairs into mutually exclusive equivalence classes. *We define a rational number to be one of these equivalence classes.*

To make the definition fruitful we must define addition, subtraction, multiplication, division and order as applied to these newly defined rationals. In each case we will define them in terms of representative pairs in the equivalence classes, and have to show the results independent of our choice of pairs. We are guided in our definitions by the known properties of fractions that we wish our rationals to possess.

Addition. Define $(a, b)+(c, d)$ to be $(ad+bc, bd)$.

We first notice that since $b \neq 0$ and $d \neq 0$ then $bd \neq 0$ and so the definition is valid. It is independent of choice within equivalence. For suppose $(a', b')R(a, b)$ and $(c', d')R(c, d)$. Then $a'b = b'a$ and $c'd = d'c$. Now

$$(a', b')+(c', d') = (a'd'+b'c', b'd')$$

and we must show that this is in the same class as $(ad+bc, bd)$, i.e. we must show that $(a'd'+b'c')bd = (ad+bc)b'd'$, which is so as $a'b = ab'$ and $c'd = cd'$.

In order for the rationals to behave like the integers we must also show that the Commutative Law holds, that $(a, b)+(c, d) = (c, d)+(a, b)$, and also the Associative Law that

$$((a, b)+(c, d))+(e, f) = (a, b)+((c, d)+(e, f)).$$

The Commutative Law is obvious, as is the Associative Law, by working out each side.

Subtraction. All pairs of the form $(0, a)$ are in the same class, and any pair equivalent to $(0, a)$ is of this form, as may be easily seen. We call this class '0' and notice that added to any class it leaves it unchanged, since $(0, a)+(c, d) = (ac, ad)$ which is equivalent to (c, d). We define $-(a, b)$ to be $(-a, b)$. Then

$$-(a, b)+(a, b) = (-a, b)+(a, b) = (0, b^2) = 0.$$

Hence negatives are defined (easily seen to be independent of choice within class) and subtraction means addition of the negative as usual.

Multiplication. This is simpler than addition. $(a, b).(c, d)$ is defined to be (ac, bd), and $bd \neq 0$.

It is independent of choice within equivalence class. For we must show, with the same notation as for addition, that $(ac, bd)R(a'c', b'd')$, i.e. that $acb'd' = a'c'bd$, which is true since $a'b = b'a$ and $c'd = d'c$.

The Commutative and Associative Laws of multiplication are immediate.

We should also prove the Distributive Law, that

$$(a, b)((c, d)+(e, f)) = (a, b).(c, d)+(a, b).(e, f),$$

but the proof of this is straightforward and is left to the reader.

Division. The pair (a, a), which is in the same class for all $a \neq 0$, acts as the unity 1 in that multiplication by it leaves a fraction unchanged (within class). The inverse of (a, b) is defined to be (b, a), being undefined when $a = 0$, so that 0 has no inverse. Division is defined as multiplication by the inverse.

Order. $(a, b) < (c, d)$ if $ad < bc$, provided b and d are positive. If b or d is negative we take members of the same classes with both second integers positive. The definition is independent of choice within class and satisfies the laws at the end of §4.2.

Our set of rationals as defined has been shown to satisfy all the usual laws of algebra. One further point remains to be considered. We wish to include integers as special cases of rationals, but at the moment this is not so, since a rational is an equivalence class of pairs of integers. The step is a simple one however—we merely identify the class containing $(a,1)$ with the integer a. It is clear from the definition that $(a, 1)+(b, 1) = (a+b, 1)$ and $(a, 1).(b, 1) = (ab, 1)$, and so no confusion arises from the identification. The subset of classes containing such pairs as $(a, 1)$ behaves exactly as the set of integers and so may be considered *to be* the set of integers, since this considered as an abstract system possesses no properties other than those of combination and order, all of which are shared by our subset.

We now have a set of rationals which may be added, subtracted, multiplied and divided and which may be ordered. We write (a, b) as a/b and notice that $a/b = \lambda a/\lambda b$, since both are in the same class (as pairs) and hence are the same rational number. All our usual techniques for applying the four rules now follow from our definitions.

In the set of rationals any non-zero number a is a factor of any other, b, since $ax = b$ always has a solution if $a \neq 0$. There are thus no primes and all numbers except 0 are units, since all are factors of 1. Thus there is no theory of prime factorisation or of H.C.F. or L.C.M.

Theorem 5.1.1. *Given any two unequal rationals $a/b < c/d$ there is at least one other, x/y, such that $a/b < x/y < c/d$.*

For consider $(a+c)/(b+d)$. (We assume b and d positive.) Then since $a/b < c/d$, $ad < bc$ and so $ab+ad < ab+bc$, i.e. $a(b+d) < b(a+c)$ and hence $a/b < (a+c)/(b+d)$. Similarly, $ad+cd < bc+cd$, $d(a+c) < c(b+d)$ and $(a+c)/(b+d) < c/d$.

Corollary. *We can find as many rationals between any two as we like.*

For take one between them, then one between the first and the one we have already found, now one between the first and the second one found, and so on. We say that there are infinitely many rationals between any two.

Although it would seem that there exist very many rationals and it is not possible to count them in their correct order (there is no 'next' rational to a given one) yet it *is* possible to enumerate them in a different order, i.e. we may label them 1, 2, 3, ... and eventually obtain any one that we like, thus laying them out in a sequence. A set with this property is said to be *denumerable.*

Theorem 5.1.2. *The set of rationals is denumerable.*

We first lay them out in the following array:

$$\begin{array}{cccccc}
0 & 1 & 2 & 3 & 4 & \cdot\ \cdot \\
0 & -1 & -2 & -3 & -4 & \cdot\ \cdot \\
\frac{0}{2} & \frac{1}{2} & \frac{2}{2} & \frac{3}{2} & \frac{4}{2} & \cdot\ \cdot \\
\frac{0}{2} & -\frac{1}{2} & -\frac{2}{2} & -\frac{3}{2} & -\frac{4}{2} & \cdot\ \cdot \\
\frac{0}{3} & \frac{1}{3} & \frac{2}{3} & \frac{3}{3} & \frac{4}{3} & \cdot\ \cdot \\
\frac{0}{3} & -\frac{1}{3} & -\frac{2}{3} & -\frac{3}{3} & -\frac{4}{3} & \cdot\ \cdot \\
\cdot & \cdot & \cdot & \cdot & \cdot & \cdot\ \cdot \\
\cdot & \cdot & \cdot & \cdot & \cdot & \cdot\ \cdot
\end{array}$$

We now proceed diagonally starting with the top left-hand one, proceeding down the next diagonal, going down and right to left, then down the third and so on, omitting any that have already been written down. Thus the first few are as follows, the ones in brackets being omitted:

$$0, 1, (0), 2, -1, (\tfrac{0}{2}), 3, -2, \tfrac{1}{2}, (\tfrac{0}{2}), 4, -3, (\tfrac{2}{2}), -\tfrac{1}{2}, (\tfrac{0}{3}), \ldots$$

By this means we finally obtain any given rational, and so the set is denumerable.

Since the rationals cannot be enumerated in their correct order we can seldom use the method of induction: to be applied it would have to be used according to one of the orders in which they *can* be counted, and such orders are usually too complicated for simple use. Induction may of course be used for properties of the denominators alone or numerators alone.

5.2. The real numbers

The rationals give us a set of numbers which, for the first time, are capable of being manipulated by all the four basic rules, and no further extension of the number system seems necessary. We still find, however, equations that cannot be solved in terms of rationals. All linear equations $ax+b = 0$ have a solution, but quadratics need not have. Thus such a simple equation as $x^2 = 2$ has no solution if we restrict ourselves to rationals. This was realised by the Greeks, who saw that the diagonal of a unit square could not be expressed as the ratio of two integers. A proof that $\sqrt{2}$ is not rational was known to the Greeks and is reproduced below. This fact, that $\sqrt{2}$ is not rational, worried the Greek geometers, who based many of their ideas on the ratio of lengths, and although irrationals were constantly used during and after the Renaissance they were not satisfactorily explained until the nineteenth century, by Dedekind and Cantor in 1872.

Theorem 5.2.1. *$\sqrt{2}$ is not rational.*

Suppose $\sqrt{2} = p/q$ where p and q are integers with no common factor. Then $2 = p^2/q^2$, i.e. $2q^2 = p^2$. Hence p^2, and therefore p, is even, and hence p^2 is divisible by 4. Thus q^2 and so q is even, and therefore p and q have the common factor 2, which is contrary to hypothesis. Hence $\sqrt{2}$ cannot be expressed in the form p/q and so is not rational.

Since the definition of irrationals belongs to analysis rather than to pure algebra we will give a brief summary only of Dedekind's method. A full account is given in chapter 1 of G. H. Hardy's *A Course in Pure Mathematics*, and the student is recommended to read this carefully. The analytical nature of irrationals may be seen from the fact that an irrational may be considered as the limit of a certain sequence: in the case of $\sqrt{2}$ such a sequence is given by 1, 1·4, 1·41, 1·414, etc., taking $\sqrt{2}$ correct to more and more decimal places.

Dedekind defined an irrational by considering a 'section' or cut of the rationals, as follows. We imagine the rationals divided into 2 subsets L and R such that: (i) every rational belongs to one and only one of L and R, (ii) L and R are non-empty, (iii) any member of L is less than any member of R. Then either L has a greatest member, R has a least member or neither case happens. If either of the first two occurs then the greatest or least member is said to correspond to the section, while in the third case we say that the section defines an irrational number.

For example, let $x \in L$ if $x^2 < 2$ and $x \in R$ if $x^2 > 2$. Then L has no greatest and R has no least member, and the section defines $\sqrt{2}$.

Addition and multiplication of sections may be easily defined, as may negatives, subtraction and division, and order (see Hardy).

If the rational and irrational numbers (known collectively as the *real* numbers) are divided into a section then no further numbers may be obtained, since in this case it can be shown that either L has a greatest or R a least member: the third case cannot occur. Thus the real numbers complete, as it were, our line.

The set of real numbers is called the *continuum*. Between any two rationals there is an infinite number of irrationals and vice versa. The four rules may be applied, as may the notion of order. There are of course no primes, since every real number except 0 is a divisor of every other real number, and every real except 0 is a unit.

Theorem 5.2.2. The set of reals is not denumerable.

Suppose we have enumerated the reals. Let the nth be written as a decimal to at least the first n places, and let its nth decimal place be a_n, which may of course be 0 if the decimal terminates. Consider the decimal whose nth place is a_n+1 if $a_n \neq 9$, and 0 if $a_n = 9$. Then this is not contained in the enumerated set, since it differs from every number of this set in at least one place. Hence the set cannot contain all real numbers.

Because of this theorem there can be no simple induction applied directly to the real numbers.

Algebraic and transcendental numbers

A number which is a root of any polynomial equation with rational coefficients is called *algebraic*. All rationals are algebraic, as are numbers such as $\sqrt{2}$, $\sqrt[3]{3}$, etc.

A non-algebraic real number is *transcendental*: examples are π and e.

5.3. Complex numbers

The irrationals complete our line, or continuum, of numbers, but we are still unable to solve many equations in terms of them, the simplest such being $x^2+1 = 0$. In order to deal with this type of equation, i.e. to give a meaning to the square root of a negative number, complex numbers were introduced during the seventeenth century, though they were not properly understood until the end of the eighteenth.

A complex number is one of the form $a+bi$, where a and b are real and i is $\sqrt{-1}$ (or more properly *one* of the square roots chosen arbitrarily). They are manipulated according to the ordinary rules of algebra, and may be added, subtracted, multiplied and divided. We must, however, sacrifice the notion of order when dealing with them, since they cannot be represented on a line, but need two dimensions for their geometrical portrayal. The representation used is known as the 'Argand Diagram': its properties may be found in any standard work, as may the ordinary properties of complex numbers themselves.

As in the case of the rationals, these numbers may be introduced in several ways. Probably the best intuitive method is by the idea of a number as an operator. We will describe briefly the algebraic method by number pairs, which is similar to our definition of rationals though rather simpler. Only the method is indicated: for the details, and for the operator method, the reader should refer to any standard work on the subject.

Definition of complex numbers

A complex number is defined to be an ordered pair of real numbers (a, b). Addition is defined by

$$(a, b)+(c, d) = (a+c, b+d).$$

It is easily seen that $(0, 0)$ is a zero and $(-a, -b)$ the negative of (a, b), and the usual rules may be verified.

We define multiplication by $(a, b).(c, d) = (ac-bd, ad+bc)$. Then $(1, 0)$ is a unity and $[a/(a^2+b^2), -b/(a^2+b^2)]$ the inverse of (a, b) if $(a, b) \neq (0, 0)$, and the rules of multiplication may be verified.

$(a, 0)$ behaves like the set of real numbers, with $(a, 0)$ in place of a, and hence we may write $(a, 0)$ as a with no fear of later contradiction.

The pair $(0, 1)$ satisfies the equation $(0, 1).(0, 1) = (-1, 0)$, i.e. $x^2 = -1$. If we write $i = (0, 1)$ then $(a, b) = a+bi$ and the definition is complete. We can easily see that manipulation is as expected, replacing i^2 by -1 whenever it occurs.

It was found necessary to define irrationals and complex numbers in order to solve polynomial equations, and we may suspect that more new numbers would be needed. This is not so. Any polynomial equation whose coefficients are either real or complex may be solved in terms of complex numbers, and in fact a polynomial of degree n has exactly n roots, some of which may be repeated. The theorem that any such polynomial equation possesses at least one complex root is usually known as the 'Fundamental Theorem of Algebra', but its proof involves the ideas of analysis, and is fairly difficult. We will not attempt to prove it here, but if its truth is assumed then the fact that a polynomial of degree n has exactly n roots follows easily (see example 2 of §4.3).

The domain of complex numbers is said to be *algebraically closed*—no new numbers are required to solve polynomial equations within it—and its tremendous importance derives from this fact. Thus if we admit complex co-ordinates in geometry then *every* line meets a conic in two points (the roots of a quadratic), a fact which is untrue in real geometry. We thus do not need to mention exceptions in our work.

5.4. Subsets of the set of complex numbers

Certain subsets have properties similar to the basic sets that we have dealt with in the previous sections. A few examples are given.

Example 1. The even numbers behave like the integers, in that they may be added, subtracted or multiplied to give an answer within the set. The odd numbers alone do not have this property. The set of all multiples of any integer n has the same property.

Example 2. All numbers of the form $a+b\sqrt{2}$, a and b rational, form a set in which all four rules apply. It could be defined like the complex numbers

$$((a, b)+(c, d) = (a+c, b+d), \quad (a, b).(c, d) = (ac+2bd, ad+bc))$$

and the inverse of (a, b) is

$$\left(\frac{a}{a^2-2b^2}, \frac{-b}{a^2-2b^2}\right).$$

Example 3. The *Gaussian integers*, complex numbers of the form $a+ib$ where a and b are integers, may be added, subtracted and multiplied but not divided.

Example 4. The complex numbers of unit modulus may be multiplied or divided, but not added or subtracted in general.

5.5. Quaternions

Although the complex numbers form an algebraically closed set, and are therefore sufficient for our ordinary work with numbers in algebra and geometry, even this domain can be extended. The first extension was made by Hamilton with his *quaternions*, about 1850. He was led to this invention by trying to find a meaning for the quotient of two three-dimensional vectors, and he expected that his discovery would prove extremely important in mechanics. Their place was, however, taken by tensors and matrices, which are more general and more useful concepts, although the use of quaternions was advocated very strongly for some time after Hamilton's death by their supporters. Today they are of theoretical interest only, as an example of a number system more extensive than the complex numbers.

Although they were first introduced in terms of vectors, we will define a quaternion as an ordered set of four real numbers (a, b, c, d) combining according to certain laws. These are:

$$(a, b, c, d)+(e, f, g, h) = (a+e, b+f, c+g, d+h)$$
$$\lambda(a, b, c, d) = (\lambda a, \lambda b, \lambda c, \lambda d),$$

where λ is a real number. Then if we denote $(1, 0, 0, 0)$ by 1, $(0, 1, 0, 0)$ by i, $(0, 0, 1, 0)$ by j and $(0, 0, 0, 1)$ by k, we have $(a, b, c, d) = a+bi+cj+dk$.

We now define multiplication by writing two quaternions in this latter form and multiply according to the usual rules writing,

$$i^2 = j^2 = k^2 = -1, ij = -ji = k, jk = -kj = i, ki = -ik = j.$$

It can be easily verified that the quaternions with these definitions satisfy all our usual laws of addition, subtraction, multiplication and division with the exception of the Commutative Law of Multiplication, for $ij \neq ji$, etc. They are not, of course, ordered.

Thus when we extend the complex numbers (pairs of reals) to the quaternions (sets of four reals) we lose the commutative property of multiplication. There is no analogue for triples of reals: the next such numbers are sets of eight reals, known as the Cayley numbers, and they lose yet more of the ordinary laws of algebra.

Worked exercises

1. Prove from the definition in §5.1 that

$$(x+y, 9x)-(x-y, 6y)-(1, 1) = (2y^2-13xy-3x^2, 18xy).$$

$$\begin{aligned}(x+y, 9x)-(x-y, 6y) &= ((x+y)(6y)-(9x)(x-y), 54xy)\\ &= (6y^2+15xy-9x^2, 54xy).\end{aligned}$$

Hence, subtracting (1, 1) we obtain

$$\begin{aligned}((6y^2+15xy-9x^2)-54xy, 54xy) &= (6y^2-39xy-9x^2, 54xy)\\ &= (2y^2-13xy-3x^2, 18xy),\end{aligned}$$

dividing both numbers by 3. The above process is, of course, the same as the usual one, merely using different notation. It does not use the refinement of finding the L.C.M. of the denominators, and we need to cancel in the above at the end. Such refinements could be developed in this notation, but the standard method of denoting division by a line is simpler in practice, used in conjunction with our ingrained habits of technique.

2. Show how to express a repeating decimal in the form p/q, thus showing that it is rational.

Let the decimal be $\cdot a_1a_2 \ldots a_m\dot{b}_1b_2 \ldots \dot{b}_n$. (We may assume it less than 1 for convenience.) Its value is

$$\begin{aligned}&\frac{a_1a_2 \ldots a_m}{10^m}+\frac{b_1b_2 \ldots b_n}{10^{m+n}}\left(1+\frac{1}{10^n}+\frac{1}{10^{2n}}+\ldots\right)\\ &= \frac{a_1a_2 \ldots a_m}{10^m}+\frac{b_1b_2 \ldots b_n}{10^{m+n}}\left(\frac{1}{1-(1/10^n)}\right)\\ &= \frac{a_1a_2 \ldots a_m}{10^m}+\frac{b_1b_2 \ldots b_n}{10^m}\left(\frac{1}{10^n-1}\right)\end{aligned}$$

and this may be expressed as a single fraction p/q.

3. Verify that for complex numbers $(a/(a^2+b^2), -b/(a^2+b^2))$ is the inverse of (a, b), if $(a, b) \neq (0, 0)$.

$$\begin{aligned}\left(\frac{a}{a^2+b^2}, -\frac{b}{a^2+b^2}\right).(a, b) &= \left(\frac{a^2}{a^2+b^2}+\frac{b^2}{a^2+b^2}, \frac{ab}{a^2+b^2}-\frac{ab}{a^2+b^2}\right)\\ &= (1, 0).\end{aligned}$$

4. If $x = 2-3i-k$, $y = 1+j-2k$ find xy, yx.

$$\begin{aligned}xy &= 2+2j-4k-3i-3k-6j-k+i-2\\ &= -2i-4j-8k,\\ yx &= 2-3i-k+2j+3k-i-4k+6j-2\\ &= -4i+8j-2k.\end{aligned}$$

Exercises 5A

1. Prove the Associative Law of addition for rationals, defined as in §5.1.

2. Prove that the negative of a rational is independent of the choice of number pair within the equivalence class.

3. Prove the Distributive Law for rationals.

4. Prove that the definition of $<$ given in §5.1 is independent of choice of number pairs within equivalence classes.

5. Prove that $<$ satisfies the laws of §4.2.

6. Prove that $(a, b) > 0$ if either $a > 0$, $b > 0$ or $a < 0$, $b < 0$, while $(a, b) < 0$ if $a > 0$, $b < 0$ or $a < 0$, $b > 0$.

Prove **7–14** from the definition in §5.1:

7. $(a, b) = (c, d) \Rightarrow (a, b) = (\lambda a + \mu c, \lambda b + \mu d)$.

8. $(a, b) = (c, d) \Rightarrow (\lambda a + \mu b, \nu a + \kappa b) = (\lambda c + \mu d, \nu c + \kappa d)$.

9. $a > b > 0 \Rightarrow (a, b) > (a+1, b+1)$.

10. $(a, b) = (a, 1) \div (b, 1) = (1, b) \div (1, a)$.

11. $(a, b) + (c, b) = (a+c, b)$.

12. $(1, 1) - (a-b, a) = (b, a)$.

13. $(2y+z, z) - (y-z, y) = (2y^2+z^2, yz)$.

14. $(1, 1) - (c+d, 2c) - (c-d, 3c) = (c-d, 6c)$.

15. Find a polynomial equation with rational coefficients which has $\sqrt{2}+\sqrt{3}$ as a root, and deduce that $\sqrt{2}+\sqrt{3}$ is algebraic.

16. Which if any of the four rules applied to the set of irrationals alone will always keep us within the set?

17. Prove that $\log_{10} 2$ is irrational.

18. Verify the Associative Law of Multiplication and the Distributive Law for complex numbers, using the definition of §5.3.

19. Prove *from the definition given in* §5.3 that, if z_1 and z_2 are complex numbers,

$$\text{(i)} \quad |z_1+z_2| \leqslant |z_1| + |z_2|,$$

$$\text{(ii)} \quad |z_1-z_2| \geqslant |z_1| - |z_2|.$$

20. Show that the inverse of the quaternion $a+bi+cj+dk$ is

$$\frac{a-bi-cj-dk}{a^2+b^2+c^2+d^2}.$$

21. If $\alpha = a+bi$ is a complex number, show that the equation

$$(x-\alpha)(x-\bar{\alpha}) = 0$$

has real coefficients, where $\bar{\alpha}$ is the conjugate $a-bi$.

22. If $\alpha = a+bi+cj+dk$ is a quaternion, show that the equation

$$(x-\alpha)(x-\bar{\alpha}) = 0$$

has real coefficients, where $\bar{\alpha}$ is the conjugate quaternion $a-bi-cj-dk$.

23. If $x = 1+2i+k$, $y = 2+i$, $z = 1-i-j$, find xy, yx, xyz, zxy, zyx.

Exercises 5B

1. Prove that if p and q are co-prime integers, $\sqrt{p/q}$ is rational if and only if both p and q are perfect squares.

2. Prove that every terminating or repeating decimal is rational, and conversely that every rational number may be expressed as either a terminating or repeating decimal. Show further that the necessary and sufficient condition for p/q to terminate is that q contains only the prime factors 2 and 5.

3. Prove by using example 2 or otherwise that there is at least one irrational and at least one rational between any two numbers, and deduce that there are infinitely many.

4. Show that the repeating part of p/q contains at most $q-1$ digits.

5. Defining an algebraic number as in §5.2, prove that any such number is the root of a polynomial equation with *integer* coefficients.

6. Prove that if a and b are co-prime positive integers then $\log_a b$ is irrational.

7. Show that complex roots of a polynomial equation with real coefficients occur in conjugate pairs; i.e. that if $x+iy$ is a root then $x-iy$ is also. Deduce that a cubic has at least one real root.

8. State and prove the necessary and sufficient conditions that two complex numbers of unit modulus may be added to give another of unit modulus.

9. Show that the quaternion $x = bi+cj+dk$ is a root of $x^2 = -1$ provided $b^2+c^2+d^2 = 1$ and therefore that the theorem that a polynomial of degree n has n roots does not hold in the set of quaternions.

How many roots has $x^2 = +1$? How many roots has $x^2 = \alpha$, where α is a non-real quaternion?

6

RESIDUE CLASSES

6.1. Congruences

One of the main purposes of this book is to show how algebraic processes, which are often thought to apply only to numbers, are in fact of much more general application and may be applied to sets of elements which are completely different from the ordinary numbers. In chapter 3 we worked with sets as a whole and introduced algebraic operations, $\cup$ and $\cap$, which behaved like the ordinary four rules in some ways, but not in all. In the present chapter we are on more familiar ground in that we are using the four rules themselves and are applying them to elements which are very like the positive integers, but which, nevertheless, differ in some important and far-reaching aspects: for example, there are only a finite number of elements in our set, and it is not always possible to cancel common factors. The work is quite easy to understand and is used extensively in the theory of numbers.

We begin by considering the division of the integers, positive or negative, into even and odd numbers. We then notice that if we add two even numbers the answer is always even, two odds added give an even, while even plus odd is odd; thus whether the answer is even or odd depends only on the evenness or oddness of the two numbers added, and similar results hold for multiplication. Thus for certain purposes (those that depend only on the properties of being even or odd) we may treat all odd numbers alike, and all even ones alike, i.e. the multiples of 2 are immaterial and the numbers 6 and 8, say, are 'equal'. Exactly the same kind of result is true for all multiples of 3, all numbers of the form $3n+1$ and all $3n+2$: here we may ignore multiples of 3. The same holds for multiples of any other positive integer.

The above leads to the idea of congruences, which we study in this section as the first step to that of residues.

Definition. *If a and b are two integers such that $a-b$ is divisible by n, we say that a is congruent to b modulo n, and write $a \equiv b \pmod{n}$, or merely $a \equiv b(n)$.*

This means of course that a and b leave the same remainder when divided by n. Note that a and b may be positive or negative, while n is usually taken as a positive integer. Each of the pair is equal to the other added to a multiple of n. If the modulus n is known, or sometimes when it is immaterial, we merely write $a \equiv b$.

Congruences may be combined according to the usual rules of addition and multiplication, except for division, as in the case of evens and odds already mentioned. This is shown by the following theorem. (Note that $b+d(n)$ means $(b+d) \pmod{n}$, etc.)

Theorem 6.1.1. *$a \equiv b(n)$ and $c \equiv d(n) \Rightarrow$*

(i) $a+c \equiv b+d(n)$;

(ii) $\lambda a+\mu c \equiv \lambda b+\mu d(n)$ for any integers λ, μ;

(iii) $ac \equiv bd(n)$.

Since $a \equiv b(n)$ we have $a = b+rn$, where r is an integer. Since $c \equiv d(n)$ we have $c = d+sn$, where s is an integer. Hence

$$a+c = b+d+(r+s)n \quad \text{and so} \quad a+c \equiv b+d(n) \quad \text{(i)},$$

$$\lambda a+\mu c = \lambda b+\mu d+(\lambda r+\mu s)n \quad \text{and} \quad \lambda a+\mu c \equiv \lambda b+\mu d(n) \text{(ii)},$$

$$ac = bd+(dr+bs+rsn)n \quad \text{and} \quad ac \equiv bd(n) \quad \text{(iii)}.$$

Corollary. *Putting $\lambda = 1$ and $\mu = -1$ in (ii) we have*

$$a-c \equiv b-d(n).$$

Division is sometimes possible and sometimes not. Thus we may divide the congruence $4 \equiv 18(7)$ by 2 and obtain the correct congruence $2 \equiv 9(7)$, while a counter example is obtained by dividing $6 \equiv 15(9)$ by 3 and obtaining the invalid result $2 \equiv 5(9)$. The conditions under which division may be performed, sometimes with a modification of the modulus, are given in the next theorem.

Theorem 6.1.2. *$ac \equiv bd(n)$ and $c \equiv d(n) \Rightarrow a \equiv b(n/h)$, where h is the H.C.F. of d and n (or of c and n since c and d differ by a multiple of n).*

$ac = bd + rn$ for some integer r, and $c = d + sn$ for some integer s. Hence $ad + asn = bd + rn$, i.e. $(a-b)d = (r-as)n$. Now let $d = hk$, where k and n/h are co-prime. Then $(a-b)k = (r-as)(n/h)$. But k and n/h are co-prime and hence k is a factor of $(r-as)$, say $r-as = km$. Thus $a-b = m(n/h)$ and the result follows.

Note why we cannot prove that $a \equiv b \pmod{n}$. We cannot deduce that d is a factor of $r-as$ and so cannot prove that $a-b$ is a multiple of n.

Corollaries. (i) *$ax \equiv bx(n) \Rightarrow a \equiv b(n/h)$ where h is the* H.C.F. *of x and n.*

(ii) *$ac \equiv bd(n)$ and $c \equiv d(n) \Rightarrow a \equiv b(n)$ if d and n are co-prime.*

(iii) *$ax \equiv bx(n) \Rightarrow a \equiv b(n)$ if x and n are co-prime, and $ax \equiv bx(n) \Rightarrow a \equiv b(n)$ for all x which are not multiples of n if n is a prime.*

Examples

1. *Example of the theorem*: $7.3 \equiv 10.12(9)$ (i.e. $21 \equiv 120(9)$), and $3 \equiv 12(9)$. Hence, since the H.C.F. of 3 and 9 is 3, $7 \equiv 10$ $(9/3 = 3)$, which is of course true.

2. *Example of corollary* (i): $2.10 \equiv 9.10(14)$, and thus $2 \equiv 9(14/2 = 7)$.

3. *Example of corollary* (ii): $3.7 \equiv 11.15(4)$ and $7 \equiv 15(4)$, while 15 and 4 are co-prime. Hence $3 \equiv 11(4)$.

4. *Example of corollary* (iii): $5.4 \equiv 26.4(7)$; hence $5 \equiv 26(7)$.

The work of the rest of the chapter depends on the theorems above and also on the next theorem, which appears fairly obvious in the present case, the proof being important in that it generalises later (see §13.6).

Theorem 6.1.3. *The relation of congruence relative to a given modulus n is an equivalence relation.*

Congruence is certainly a relation between pairs of integers.

Reflexive. $x \equiv x(n)$ since $x - x = 0$ and so is divisible by n.

Symmetric. If $x \equiv y$ we have $x-y$ divisible by n and so $y-x$ is.

Transitive. If $x \equiv y$ and $y \equiv z$ we have $x-y = rn$, $y-z = sn$ and so $x-z = (r+s)n$, giving $x \equiv z$.

6.2. Residue classes

It follows from theorem 6.1.3 that the relation of congruence modulo n, being an equivalence relation, decomposes the set of integers into mutually exclusive equivalence classes, such that two integers are in the same class if and only if they are congruent modulo n. These classes are called *residue classes* modulo n. We can easily see that there are just n residue classes modulo n, a typical one consisting of all those integers that leave remainder r when divided by n, where r ranges over the values $0, 1, \ldots, (n-1)$. The class of the integers congruent to any of these r's is often denoted merely by r itself, when it is understood that we are working with residues. To save confusion in this chapter we will use bold type for residue classes: thus $\mathbf{r}$ means the residue class of integers congruent to r, i.e. those leaving remainder (or 'residue') r when divided by n. In future chapters we revert to the usual notation and relinquish the bold type. We will sometimes shorten the term 'residue class' to 'residue'.

Examples of residue classes

Example 1. If $n = 2$ the equivalence relation of congruence divides the integers into two residue classes, one comprising all even numbers, denoted by $\mathbf{0}$, and the other, denoted by $\mathbf{1}$, comprising all odd integers.

Example 2. Instead of writing the numbers along a line, as in the continuum, let us rather put them round a circle, starting at 0 and writing each successive positive integer at an angular distance of $2\pi/n$ from its predecessor. (For $n = 12$ the resulting diagram would resemble a clock face.) We now see that n is at the same place as 0, $n+1$ is with 1, and so on. The residue class $\mathbf{r}$ consists of all those integers at the place r.

Example 3. Consider the set C consisting of all multiples of n. Then the residue class $\mathbf{r}$ is obtained by adding r to each member of C in turn. Thus in some sense the set of elements in $\mathbf{r}$ is 'parallel' to C. (Note that $\mathbf{0}$ is C itself.) It lacks one important property of C in that the sum or product of two members of $\mathbf{r}$ is not necessarily itself in $\mathbf{r}$ (the sum never is unless $r = 0$).

Example 4. Consider all positive integers whose last digit is

r. These consist of all the positive members of $\mathbf{r}$, modulo 10. A similar result holds modulo n if we use the n-ary scale of notation.

The algebra of residues

We have already said that if we add two evens we obtain an even, with similar results for two odds or an even and an odd; the quality of evenness or oddness depends only on whether the summands are even or odd, the same being true for multiplication. Thus we may give a meaning to addition or multiplication of two residue classes modulo 2. The same is true for any modulus.

Theorem 6.2.1. *If we add any member of* $\mathbf{r}$ *to any member of* $\mathbf{s}$ *we obtain a member of the class containing* $r+s$.

For $(r+an)+(s+bn) = (r+s)+(a+b)n$.

Note. (1) We cannot say that we obtain a member of $\mathbf{u}$ (where $u=r+s$) since $r+s$ may be greater than n, and we are denoting a class by its *smallest* positive member. The important thing is that all the sums are in the same residue class, which is of course $\mathbf{t}$, where $t \equiv r+s(n)$ and $0 \leqslant t \leqslant n-1$.

(2) The theorem is a direct consequence of theorem 6.1.1.(i).

The above theorem enables us to define addition of two residue classes. We define $\mathbf{r}+\mathbf{s}$ to be that class, uniquely determined because of the theorem, which contains all sums of pairs of integers, one from $\mathbf{r}$ and one from $\mathbf{s}$. $\mathbf{r}+\mathbf{s}$ is thus the class that contains $r+s$ and is obtained by subtracting the necessary multiples of n from $r+s$ to obtain a remainder less than n. Thus modulo 2 we have $\mathbf{1}+\mathbf{1} = \mathbf{0}$ (odd plus odd is even), while $\mathbf{1}+\mathbf{0} = \mathbf{1}$ (odd plus even is odd).

This idea is vitally important. We are using a process of addition with a set that, despite its superficial resemblance to the set of integers, is really completely different. It has only a finite number (n) of elements, and is cyclic in that after adding 1 for a finite number of times we reach 0 again. Yet the addition we have defined is easily seen to possess all the properties that we associate with the usual addition of integers. It is commutative ($\mathbf{r}+\mathbf{s} = \mathbf{s}+\mathbf{r}$), associative ($\mathbf{r}+(\mathbf{s}+\mathbf{t}) = (\mathbf{r}+\mathbf{s})+\mathbf{t}$), there is an element **0** which does not alter another element when added to it, and we will see below that negatives have a meaning. We have a refutation of the popular idea that 'one plus one *always* equals two', for in the set of residue classes modulo two it equals zero. The idea is useful, as well as being theoretically important, for not only is it vital in the theory of numbers, but there are some branches of mathematics where the elements we use are not important—it is merely required that they can be added and possibly multiplied. In this case we may use the residues modulo 2 (say) and thus have only two elements in our set, making the work extremely simple arithmetically.

Theorem 6.2.2. (i) $\mathbf{r}+(\mathbf{n}-\mathbf{r}) = \mathbf{0}$.

(ii) *If from any member of* **r** *we subtract any member of* **s** *we obtain a member of the class containing* $r-s$.

The proofs are obvious.

We define $\mathbf{n}-\mathbf{r}$ to be the class $-\mathbf{r}$. This has the basic property of negatives that a negative added to its positive gives zero.

The class containing $r-s$ is defined as the difference $\mathbf{r}-\mathbf{s}$.

Theorem 6.2.3. *If we multiply any member of* **r** *by any member of* **s** *we obtain a member of the class containing* rs.

This follows from theorem 6.1.1.(iii), or may be proved directly.

We define **rs** to be the class containing rs, and by the theorem this gives a proper meaning to multiplication of residues, which satisfies the Commutative and Associative Laws of multiplication, and also the Distributive Law

$$\mathbf{r}(\mathbf{s}+\mathbf{t}) = \mathbf{rs}+\mathbf{rt}.$$

The class **1** acts as a 'neutral' element in that $\mathbf{1r} = \mathbf{r}$.

For any positive integer n we now possess a set of n elements within which we can add, subtract, or multiply, just as with the set of integers. There are two major differences. First, there is no meaningful definition of order. At first sight we would think that we could say $\mathbf{r} < \mathbf{s}$ if $r < s$, but a little examination shows us that this definition does not satisfy such basic laws as $\mathbf{a} < \mathbf{b}$ and $\mathbf{c} < \mathbf{d} \Rightarrow \mathbf{a}+\mathbf{c} < \mathbf{b}+\mathbf{d}$ (for example, modulo 10 we have $\mathbf{2} < \mathbf{6}$ and $\mathbf{3} < \mathbf{5}$, but $\mathbf{5} \nless \mathbf{1}$), and has therefore no practical use. Although we select the member of our class which lies between 0 and $n-1$ as a label for the whole class, in fact no member is of more importance than any other, thus depriving the notion of $<$ of its meaning, since by judicious choice of the members we could have *any* residue class 'less than' any other, or indeed 'greater than'.

The second difference between the sets of residue classes and the set of integers is the fact that the Cancellation Law of multiplication does not necessarily hold with residues. Thus for integers $ax = bx$ necessarily implies that $a = b$ unless $x = 0$, and this further implies that $ab = 0 \Rightarrow a = 0$ or $b = 0$. For residues this need not be so. Thus $\mathbf{4}.\mathbf{5} = \mathbf{6}.\mathbf{5}$ modulo 10 (each $= \mathbf{0}$) but $\mathbf{4} \neq \mathbf{6}$. Also neither $\mathbf{4}$ nor $\mathbf{5} = \mathbf{0}$ but $\mathbf{4}.\mathbf{5}$ does. The Cancellation Law is closely connected with the possibility or otherwise of division, which forms the subject of the next section.

6.3. Division of residues

We have seen that the Cancellation Law does not necessarily hold for residues. Neither is division of one residue by another always possible. Before we investigate this question we need a useful definition of division in this context. We cannot use the obvious definition, considering all quotients a/b, where a is any member of $\mathbf{r}$ and b any member of $\mathbf{s}$, since this quotient may not always be an integer, and even when it is it will not always be the same integer. A better definition is in terms of multiplication. With numbers, a/b means the number which when multiplied by b gives a, i.e. the solution of $bx = a$. So with residues we will define $\mathbf{s}/\mathbf{r}$ to be a solution of $\mathbf{rx} = \mathbf{s}$. Then for division to exist such an $\mathbf{x}$ must exist and be unique. This may or may not be so.

Examples. Modulo 4 there is a unique solution of $\mathbf{3x} = \mathbf{2}$, viz. $\mathbf{2}$. There is no solution of $\mathbf{2x} = \mathbf{3}$, while both $\mathbf{1}$ and $\mathbf{3}$ are solutions of $\mathbf{2x} = \mathbf{2}$.

The circumstances under which we can divide residues are given in the next theorem.

Theorem 6.3.1. *If $r \neq 0$, division by $\mathbf{r}$, in the residue classes modulo n, is possible if and only if r is prime to n.*

If r is prime to n. By theorem 4.5.3. (corollary) there exist integers a and b such that $ra+nb = 1$. Hence for any s we have $r.as+n.bs = s$, i.e. $r.as \equiv s \pmod{n}$. Thus $\mathbf{r}.\mathbf{as} = \mathbf{s}$ and so $\mathbf{as}$ is a solution of $\mathbf{rx} = \mathbf{s}$. We must show that it is the unique solution. We have proved that $\mathbf{rx} = \mathbf{s}$ has a solution for all $\mathbf{s}$, and so as $\mathbf{x}$ moves over the range $\mathbf{0}, \mathbf{1}, \ldots, (\mathbf{n}-\mathbf{1})$, $\mathbf{rx}$ must take each of the values $\mathbf{0}, \mathbf{1}, \ldots, (\mathbf{n}-\mathbf{1})$ at least once, and therefore exactly once. Hence the solution is unique.

Notice that the solution of $\mathbf{rx} = \mathbf{1}$ is $\mathbf{a}$ (the 'inverse' of $\mathbf{r}$) and the solution of $\mathbf{rx} = \mathbf{s}$ is then $\mathbf{as}$, i.e. $\mathbf{s}/\mathbf{r} = \mathbf{sr}^{-1}$, as is the case for ordinary real numbers.

Since $\mathbf{rx}$ takes different values for different values of $\mathbf{x}$, the Cancellation Law holds, that $\mathbf{rx} = \mathbf{ry} \Rightarrow \mathbf{x} = \mathbf{y}$, if $\mathbf{r} \neq \mathbf{0}$. Also if $\mathbf{r}$ and $\mathbf{x}$ are neither of them $\mathbf{0}$, nor is $\mathbf{rx}$.

If r is not prime to n. Let $(r, n) = h$, we see that rx is divisible by h for all x, and so, since h is a factor of n, $\mathbf{rx} = \mathbf{y}$ where y is a multiple of h. Hence $\mathbf{rx} = \mathbf{s}$ has no solution if s is not divisible by h.

If s is a multiple of h, say $s = th$, then by theorem 4.5.3 we can find a and b such that $ra+nb = h$, and so $r.at \equiv th = s$ and $\mathbf{r}.\mathbf{at} = \mathbf{s}$, so that $\mathbf{at}$ is one solution of $\mathbf{rx} = \mathbf{s}$. If $n/h = k$, $rk = n(r/h) \equiv 0 \pmod{n}$ since r/h is an integer. Hence

$$\mathbf{r}(\mathbf{at}+\lambda\mathbf{k}) = \mathbf{s} \quad \text{for} \quad \lambda = 0, 1, \ldots, h$$

and so $\mathbf{at}+\lambda\mathbf{k}$ is a solution of $\mathbf{rx} = \mathbf{s}$, there is no unique solution and so division is not possible.

The Cancellation Law doesn't hold, since we can deduce only that $\mathbf{rx} = \mathbf{ry} \Rightarrow \mathbf{x} \equiv \mathbf{y} \pmod{n/h}$ by theorem 6.1.2, i.e. $\mathbf{x} = \mathbf{y} + \lambda\mathbf{k}$ as before.

The arithmetic modulo p, where p is prime

If n is a prime p, then *all* numbers which are not multiples of p are prime to p and division by non-zero elements is always possible. We thus have a set of p elements within which all four rules, addition, subtraction, multiplication, and division, may be carried out. We have an algebraic structure more complete even than is present in the integers—we need to include rationals to make division possible there—and we are working with a finite number of elements, which gives a system intrinsically simpler than the usual infinite ones. The system is often called the *arithmetic modulo p*, and we work with its elements according to the usual rules, but omitting all multiples of p. In practice it is usual to write the elements $\mathbf{r}$ like the number r, the fact that we are working modulo p being understood in any piece of work. Some of these finite arithmetics will be studied in detail in §6.4, and mentioned in later chapters of this and the subsequent volume.

The arithmetic modulo n, where n is composite

If n is composite, we have a finite set within which we can add, subtract and multiply, but cannot divide in general. We have seen that division is possible for all divisors prime to n, and we note that if two such are multiplied the product is also prime to n. Hence the residues prime to n may be multiplied and divided within their set, but not of course always added (e.g. $3+3 \equiv 2(4)$, and 2 is not prime to 4, though 3 is). Such structures as the arithmetics modulo n have most of the properties of the set of integers, with the very important exception of the Cancellation Law. Their additive properties are as extensive, however, and these sets will also appear later.

Within these finite arithmetics there is no theory of primes or of divisibility. In the prime modulus case every element is a divisor of every other, while for composite moduli there are still no primes, since every element is divisible by at least all the residues prime to n.

6.4. Addition and multiplication tables: some particular finite arithmetics

In particular cases we can see quickly how the finite arithmetics behave by constructing tables for addition and multiplication, in an obvious manner which we use in considering some of the smallest moduli.

The arithmetic modulo 2

	0	1
0	0	1
1	1	0

Addition

	0	1
0	0	0
1	0	1

Multiplication

This is the simplest of all arithmetics, consisting merely of the numbers 0 and 1. The tables for addition and multiplication are shown above. We see that they correspond to the rules of combination of even and odd numbers if 0 represents even and 1 odd. The addition table also corresponds to the rules for *multiplying* positive and negative numbers together, if 0 represents positives and 1 negatives: viz. the product of two pluses or two minuses is plus, while that of a plus and a minus is minus.

Since 2 is prime we can perform all four rules in this arithmetic. Since $-1 = +1$ and of course $-0 = +0$ we see that there is no such thing as mistake in sign when working with it: this fact is useful in some advanced branches of mathematics, such as topology, where our choice of number system is often left at our disposal.

The arithmetic modulo 5

	0	1	2	3	4
0	0	1	2	3	4
1	1	2	3	4	0
2	2	3	4	0	1
3	3	4	0	1	2
4	4	0	1	2	3

Addition

	0	1	2	3	4
0	0	0	0	0	0
1	0	1	2	3	4
2	0	2	4	1	3
3	0	3	1	4	2
4	0	4	3	2	1

Multiplication

The multiplication table is the more interesting. Note that each number occurs once and only once in each row or column except the first: this corresponds to the Cancellation Law, which holds here since 5 is prime. The same property holds for the addition table, where each row is a cyclic permutation of the one above.

The inverses are given by the position of the 1 in a row—the inverses of 1, 2, 3, 4 are respectively 1, 3, 2, 4 and so 1 and 4 are their own inverses. The values of r^2 are given by the diagonal, so we see that the equations $x^2 = 1$ and $x^2 = 4$ have two solutions each, while $x^2 = 2$ and $x^2 = 3$ have no solutions.

The arithmetic modulo 4

	0	1	2	3
0	0	1	2	3
1	1	2	3	0
2	2	3	0	1
3	3	0	1	2

Addition

	0	1	2	3
0	0	0	0	0
1	0	1	2	3
2	0	2	0	2
3	0	3	2	1

Multiplication

Although the addition table still has one of each number in every row and column, corresponding to the Cancellation Law of addition, the multiplication table does not. The contents of a row are all different for the rows corresponding to 1 and 3, but not for that corresponding to 2, i.e. only those rows corresponding to numbers prime to 4, as was proved in theorem 6.3.1. The row corresponding to 2 contains 2 and 0 twice each, since these are those numbers less than 4 which are divisible by 2. Thus 2 has no inverse, while 1 and 3 are their own inverses.

Note that if we take that part of the multiplication table modulo 5 which misses out the row and column corresponding to 0, and replace the numbers 1, 2, 3, 4 by 0, 1, 3, 2, respectively, then after rearranging rows and columns we have the addition table modulo 4. A similar result is obtained by replacing 1, 2, 3, 4 by 0, 3, 1, 2, respectively. Thus in a sense the truncated multiplication table modulo 5 is the same as the addition table modulo 4: such tables are called 'isomorphic', a word that will be found very important later.

6.5. Congruence equations

An equation $f(x) \equiv 0$ (modulo some n) is called a *congruence equation*. We will investigate briefly some polynomial congruence equations. We may write one such as $f(\mathbf{x}) = \mathbf{0}$, using residues mod n, and its form will be

$$f(\mathbf{x}) = \mathbf{a}_m\mathbf{x}^m + \mathbf{a}_{m-1}\mathbf{x}^{m-1} + \ldots + \mathbf{a}_1\mathbf{x} + \mathbf{a}_0 = \mathbf{0}.$$

If $\mathbf{a}_m \neq \mathbf{0}$ the equation has degree m, and a value of $\mathbf{x}$ which satisfies it is called a *root* of the equation.

A polynomial equation in ordinary numbers of degree n has exactly n roots, if we include complex roots, but this is by no means so for a congruence equation. Thus the general equation of the first degree, $\mathbf{ax} = \mathbf{b}$, has a unique root if a is prime to n (theorem 6.3.1); while if a is not prime to n it may have no root, or more than one root. Similar complications occur for congruence equations of higher degree: $f(\mathbf{x}) = \mathbf{b}$, as b varies, can have a *total* number of only n distinct roots, since there are only n different values of x, and so it is clearly unlikely for each of these equations to have m roots if m, the degree of $f(\mathbf{x})$, is greater than one, such an event requiring repeated roots for all the equations. For a prime modulus we may restrict the roots by the following theorem.

Theorem 6.5.1. *A congruence equation of degree m, with a prime modulus p, cannot have more than m roots, though of course it need not always have m.*

Let the equation be $\mathbf{a}_m\mathbf{x}^m + \mathbf{a}_{m-1}\mathbf{x}^{m-1} + \ldots + \mathbf{a}_0 = \mathbf{0}$, and suppose $\mathbf{x}_1$ is one root. Then $\mathbf{a}_m\mathbf{x}_1^m + \mathbf{a}_{m-1}\mathbf{x}_1^{m-1} + \ldots + \mathbf{a}_0 = \mathbf{0}$ and any other root $\mathbf{x}$ satisfies

$$\mathbf{a}_m(\mathbf{x}^m - \mathbf{x}_1^m) + \mathbf{a}_{m-1}(\mathbf{x}^{m-1} - \mathbf{x}_1^{m-1}) + \ldots + \mathbf{a}_1(\mathbf{x} - \mathbf{x}_1) = \mathbf{0}.$$

But

$$\mathbf{x}^r - \mathbf{x}_1^r = (\mathbf{x}^{r-1} + \mathbf{x}_1\mathbf{x}^{r-2} + \mathbf{x}_1^2\mathbf{x}^{r-3} + \ldots + \mathbf{x}_1^{r-1})(\mathbf{x} - \mathbf{x}_1),$$

and so $(\mathbf{x} - \mathbf{x}_1)(\mathbf{a}_m\mathbf{x}^{m-1} + \mathbf{b}_{m-2}\mathbf{x}^{m-2} + \ldots + \mathbf{b}_0) = \mathbf{0}$ for some $\mathbf{b}_{m-2}, \mathbf{b}_{m-3}, \ldots, \mathbf{b}_0$. Thus since we are working with a prime modulus, and $\mathbf{x} - \mathbf{x}_1 \neq \mathbf{0}$ we have that any root other than $\mathbf{x}_1$ is a root of $\mathbf{a}_m\mathbf{x}^{m-1} + \mathbf{b}_{m-2}\mathbf{x}^{m-2} + \ldots + \mathbf{b}_0 = \mathbf{0}$, a polynomial of degree $m-1$. Now use induction and assume that any congruence equation of degree $m-1$ has at most $m-1$ roots. Then the result follows for equations of degree m, and it is clearly true for those of degree one, since p is prime.

This theorem gives an upper limit for the number of roots, but does not give any lower limit—there may well be no roots at all. For example, the equation $x^2 \equiv 2$ (modulo 3) has no solutions.

Theorem 6.5.2. *The equation* $\mathbf{x}^2 = \mathbf{a}$, *for a prime modulus* $p > 2$, *has either no roots or two roots* $\mathbf{r}$ *and* $\mathbf{p} - \mathbf{r}$, *except that* $\mathbf{x}^2 = \mathbf{0}$ *has the one root* $\mathbf{0}$ *only.*

Consider the equation $\mathbf{x}^2 = \mathbf{r}^2$, i.e. $(\mathbf{x}-\mathbf{r})(\mathbf{x}+\mathbf{r}) = \mathbf{0}$. Then one factor is $\mathbf{0}$ since we are working with a prime modulus, and so the two solutions are $\mathbf{r}$ and $-\mathbf{r}$, i.e. $\mathbf{p}-\mathbf{r}$. Thus if the equation has one root it has two, unless of course $\mathbf{r} = \mathbf{0}$.

Corollary. *There are exactly* $\frac{1}{2}(p-1)$ *different values of* $\mathbf{x}^2$, *excluding the value* $\mathbf{0}$. *(p is odd since it is a prime* > 2.)

If the modulus n is composite then $\mathbf{x}^2$ and $(\mathbf{n}-\mathbf{x})^2$ have the same value but this value may occur for other values of $\mathbf{x}$ as well. Thus $x^2 \equiv 1$ modulo 8 has the solutions $x = 1, 3, 5, 7$.

6.6. Some results in the theory of numbers

Theorem 6.6.1. *Fermat's theorem.*

If p is prime, $a^p \equiv a(p)$, *and if a is not a multiple of p,* $a^{p-1} \equiv 1(p)$.

The numbers $a, 2a, 3a, \ldots, (p-1)a$ are all different modulo p if a is not a multiple of p, and so they must be $1, 2, \ldots, (p-1)$ modulo p in some order. Thus

$$a.2a.3a\ldots(p-1)a \equiv 1.2.3\ldots(p-1).$$

Hence

$[1.2.3\ldots(p-1)]a^{p-1} \equiv 1.2.3\ldots(p-1)$, and $1.2.3\ldots(p-1)$ and p have no common factor since p is prime. Hence we may divide both sides by $1.2.3\ldots(p-1)$ and obtain $a^{p-1} \equiv 1(p)$. Thus $a^p \equiv a(p)$ if a is not a multiple of p, and this latter result is true also if a *is* a multiple of p, since then both sides are 0. Hence it is true for all a.

Theorem 6.6.2. *Euler's extension of Fermat's theorem.*

If n is not necessarily prime, and $\phi(n)$ *is the number of integers less than n and prime to n, then* $a^{\phi(n)} \equiv 1(n)$ *if a is prime to n.*

If $r_1, r_2, \ldots, r_{\phi(n)}$ are the integers less than n and prime to n, $ar_1, ar_2, \ldots, ar_{\phi(n)}$ are all distinct modulo n and are all prime to n, hence they are $r_1, r_2, \ldots, r_{\phi(n)}$ in a different order. The proof is then as in Fermat's theorem.

Theorem 6.6.3. *Wilson's theorem.*

$(p-1)! \equiv -1(p)$ *if and only if p is prime.*

If p is prime then by Fermat's theorem the congruence equation $x^{p-1}-1 \equiv 0$ is satisfied by $x = 1, 2, \ldots, (p-1)$. Hence the equation $(x-1)(x-2)\ldots(x-p+1)-(x^{p-1}-1) \equiv 0$

is also satisfied by the $p-1$ values $1, 2, \ldots, p-1$. But this second congruence has degree $p-2$ and so must be an identity by theorem 6.5.1. Hence $x = 0$ satisfies it, and thus

$$(-1)^{p-1}(p-1)!+1 \equiv 0.$$

But p is a prime and so $p-1$ is even unless $p = 2$. The result follows unless $p = 2$, and is trivially true for $p = 2$ also.

If p is not prime $(p-1)!$ and p have a common factor, and thus $(p-1)!+1$ cannot be divisible by p.

6.7. Divisibility tests

The following tests for divisibility by various integers are well known, but the proofs of some of them may not be familiar to the reader. The basis of all the proofs is the idea of congruence.

Theorem 6.7.1. *An integer n is divisible by 5 if and only if the last digit is divisible by 5 (i.e. is 5 or 0) and, more generally, n is divisible by 5^k if and only if the number formed by the last k digits is divisible by 5^k.*

If the last digit is a, then $n = 10b+a$ and so $n \equiv a \pmod 5$. Hence the condition is $a \equiv 0(5)$.

If the last k digits form the number a, $n = b.10^k+a$ and so $n \equiv a(5^k)$.

Theorem 6.7.2. *n is divisible by 2^k if and only if the number formed by the last k digits is divisible by 2^k.*

The proof is as in theorem 6.7.1.

Theorem 6.7.3. *n is divisible by 3 or 9 if the sum of its digits is divisible by 3 or 9, respectively.*

Let $n = ab \ldots k$. Then

$$n = a.10^{s-1}+b.10^{s-2}+ \ldots +j.10+k,$$

where s is the number of digits. Thus

$$n = (a+b+ \ldots +k)+(10^{s-1}-1)a+ \ldots +(10-1)j.$$

But $10^{\alpha}-1 \equiv 0(9)$ for $\alpha \geqslant 1$, since $10-1 = 9$ is a factor. Hence $n \equiv (a+b+ \ldots +k)$ modulo 9 or modulo 3 and the result follows.

Theorem 6.7.4. *n is divisible by* 11 *if the sum of the even placed digits differs from the sum of the odd placed ones by a multiple of* 11.

If

$$n = ab \dots k, \quad n = k-j+i- \dots$$
$$+(10+1)j+(10^2-1)i + (10^3+1)h+ \dots$$

and $10^{2\alpha+1}+1$ and $10^{2\alpha}-1$ are both divisible by $10+1 = 11$. Hence $n \equiv k-j+i - \dots \pmod{11}$ and the result follows.

The above theorems used in combination provide tests for divisibility by all the integers up to 20, except for 7, 13, 14, 17 and 19 (for example, a number is divisible by 15 if it is divisible by both 3 and 5). No such simple tests exist for these remaining numbers, but the following theorem gives us a method of simplifying the search for them. It applies to 7, 13, 17, 19, and divisibility by 14 obviously depends on that by 7 and 2.

Theorem 6.7.5. *If n is expressed in the form* $10a+b$, *then*

$$n \text{ is divisible by } \begin{Bmatrix} 7 \\ 13 \\ 17 \\ 19 \end{Bmatrix} \text{ if and only if } \begin{Bmatrix} a-2b \\ a+4b \\ a-5b \\ a+2b \end{Bmatrix} \text{ is}$$

For $$10a+b-3(a-2b) = 7a+7b \equiv 0(7).$$

Thus $$10a+b \equiv 0 \Leftrightarrow 3(a-2b) \equiv 0$$
$$\Leftrightarrow a-2b \equiv 0.$$

The other results follow similarly, the relevant congruences being:

$$10a+b+3(a+4b) = 13a+13b \equiv 0(13),$$
$$10a+b+7(a-5b) = 17a-34b \equiv 0(17),$$
$$10a+b+9(a+2b) = 19a+19b \equiv 0(19).$$

As an example of the above theorem in use, let us discover whether 668,819 is divisible by 19. Here $a = 66{,}881$ and $b = 9$, and so $a+2b = 66{,}899$, and 668,819 is divisible by 19 if and only if 66,899 is. We repeat the process with 66,899. Here $a = 6689$ and $b = 9$, giving $a+2b = 6707$, and repeating again gives $670+14 = 684$. A final repetition gives 76, which *is* divisible by 19, and so the original number is.

Digital root

If we add the digits of a number n to form an integer n_1, and repeat the process until we obtain a single digit (i.e. an integer between 1 and 9) this final digit is known as the *digital root* of n.

For example, if $n = 357,645,987$, $n_1 = 54$, the next integer $n_1 = 5+4 = 9$ and the digital root is 9.

Theorem 6.7.6. *The digital root is the residue of n modulo* 9 (*but is* 9 *if the residue is* 0). *Hence if two integers are added, subtracted or multiplied the digital root of the result is the sum, difference or product respectively of the digital roots of the two integers, provided we take digital roots whenever we obtain a number greater than* 9.

By the proof of theorem 6.7.3, $n \equiv n_1(9)$ and so $n \equiv$ the digital root modulo 9. But the root is between 1 and 9 and so it is the residue of n (or 9). The rest of the theorem is immediate by the theory of congruences (theorem 6.1.1).

Note. The idea of digital root provides a useful check when manipulating large integers. Thus if we multiply 456 by 328 and obtain 149,468 we check the digital roots and discover that the first two are 6 and 4: their product is 24 and the root of this is 6. But the root of 149,468 is 5 and so our answer is wrong. A check reveals that the correct product is 149,568, whose digital root is 6 as required.

Worked exercises

1. Prove that: $3^{4n+2}+5^{2n+1} \equiv 0(14)$,

$$\begin{aligned} 3^{4n+2}+5^{2n+1} &= 9.81^n+5.25^n \\ &\equiv 9.11^n+5.11^n \\ &= 14.11^n \\ &\equiv 0. \end{aligned}$$

2. Solve $39x \equiv 17(67)$ (i.e. find $17 \div 39 (\text{mod } 67)$).

Since 39 is prime to 67 we know that a unique solution exists. In theorem 6.3.1 we showed how to find the solution by determining the integers a and b which give $ra+nb = 1$, and this determination depends in its turn on Euclid's algorithm. Finding the solution in numerical examples is therefore best done in general by following the algorithmic process as shown below, although there may be instances where shorter methods will give us the result, by intuition or calculation.

Euclid's algorithm	*Solution of congruence*
$67 = 1.39+28$	$28x \equiv -39x \equiv -17$
$39 = 1.28+11$	$11x = 39x-28x \equiv 17+17 \equiv 34$
$28 = 2.11+6$	$6x = 28x-2.11x \equiv -17-2.34 \equiv 49$
$11 = 1.6+5$	$5x = 11x-6x \equiv 34-49 \equiv 52$
$6 = 1.5+1$	$x = 6x-5x \equiv 49-52 \equiv 64$
$5 = 5.1$	

Hence the solution is 64 (modulo 67, of course).

3. Solve $15x \equiv 27(39)$.

Since 15 is not prime to 39 there is no unique solution but, since 27 is a multiple of $(15, 39) = 3$, multiple solutions exist by theorem 6.3.1, three in number differing by $39/3 = 13$. As before we use Euclid's algorithm.

Euclid's algorithm	*Solution of congruence*
$39 = 2.15+9$	$9x \equiv -2.15x \equiv -54 \equiv 24$
$15 = 1.9+6$	$6x = 15x-9x \equiv 27-24 \equiv 3$
$9 = 1.6+3$	$3x = 9x-6x \equiv 24-3 \equiv 21$
$6 = 2.3$	

Hence $x \equiv 7, 20$ or 33.

4. Solve $3x^2+7x \equiv 9(11)$.

The method here is to adjust the coefficients of x and the constant term so that every term is divisible by 3.

$$3x^2+7x-9 \equiv 3x^2+18x-9 \equiv 0(11).$$

But 3 is prime to 11, and so $x^2+6x-3 \equiv 0$. Now complete the square. $(x+3)^2 \equiv 12 \equiv 1$. Hence $x+3 \equiv 1$ or 10 and $x \equiv 9$ or 7.

Exercises 6A

1. Give the addition and multiplication tables for the arithmetics modulo: (i) 3, (ii) 6, (iii) 7.

2. Find a way of interchanging the numbers 1 and 2 so that the multiplication table modulo 3 (omitting the row and column corresponding to 0) is the same as the addition table modulo 2.

3. Repeat example 2 for the multiplication table modulo 7 and the addition table modulo 6.

4. Why cannot we repeat example 2 for the multiplication table modulo 6 and the addition table modulo 5?

5. If r is prime to n prove that, in the set of residues modulo n, $\mathbf{a}$, $\mathbf{a}+\mathbf{r}$, $\mathbf{a}+2\mathbf{r}$, ..., $\mathbf{a}+(n-1)\mathbf{r}$ are all different and hence that they form a permutation of $\mathbf{0}, \mathbf{1}, \ldots, (\mathbf{n}-\mathbf{1})$.

6. Prove that a perfect square must end in one of the digits 0, 1, 4, 5, 6, 9.

7. Prove that a perfect cube may end in any digit.

8. Prove that a fourth power must end in 0, 1, 5 or 6 and deduce that a number of the form a^{2^r} where $r \geqslant 2$ must end in one of the same set of digits.

9. If p is prime prove that $(a+b)^p \equiv a^p+b^p(p)$.

10. Show that the product of r consecutive positive integers is divisible by $r!$

11. Show that $n(n+1)(2n+1) \equiv 0(6)$.

12. Show that if n is odd, $(n^2+3)(n^2+7) \equiv 0(32)$.

13. Show that $13^{2n}-1 \equiv 0(168)$.

14. If a and b are both prime to p, where p is itself prime, prove that $a^{p-1}-b^{p-1} \equiv 0(p)$.

15. Prove that $n^5-n \equiv 0(30)$.

16. Prove that for all a, $a^{12} \equiv 0$ or $1(13)$.

17. Show that for all a, $a^3 \equiv 0$ or $\pm\ 1(7)$.

18. Prove that $a^2+b^2 \equiv 0(3) \Rightarrow a \equiv 0(3)$ and $b \equiv 0(3)$.

19. Prove that the sum of the integers less than and prime to n is $\frac{1}{2}n\phi(n)$.

20. Prove that $3.5^{2n+1}+2^{3n+1} \equiv 0(17)$.

Solve the congruence equations in **21–30**, giving all solutions in cases where the modulus and divisor are not co-prime. If there is no solution say so.

21. $3x \equiv 1(7)$.

22. $6x \equiv 4(11)$.

23. $17x \equiv 23(29)$.

24. $7x \equiv 39(50)$.

25. $75x \equiv 1(343)$.

26. $15x \equiv 3(27)$.

27. $15x \equiv 4(27)$.

28. $30x \equiv 75(145)$.

29. $15x \equiv 19(25)$.

30. $113x \equiv 226(339)$.

31. Solve $x^2+4x \equiv 22(23)$.

32. Solve $7x^2-3x \equiv 2(29)$.

33. Find values of $\mathbf{a}$ ($\neq \mathbf{0}$ or $\mathbf{1}$) such that $\mathbf{a}^2 = \mathbf{a}$: (i) mod 6, (ii) mod 10.

Exercises 6B

1. Prove the converse of theorem 6.1.2 and its corollaries; i.e. that $c \equiv d(n)$ and $a \equiv b(n/h) \Rightarrow ac \equiv bd(n)$, where $h = (d, n)$, etc.

2. If p is prime give a direct proof of the Cancellation Law for residues modulo p and deduce that division is always possible.

3. If p is prime, $\mathbf{a}^{p-1} = \mathbf{1}$ in the set of residues modulo p, by Fermat's Theorem. $\mathbf{a}$ is called a *primitive* $(p-1)$th root of unity if no lower power of $\mathbf{a}$ is equal to $\mathbf{1}$. Find all the primitive roots when $p = 3, 5, 7, 11$, respectively.

4. If $\phi(n)$ is the number of integers less than and prime to n, prove that if $(a, b) = 1$, $\phi(ab) = \phi(a).\phi(b)$.

(*Hint.* Write the numbers $1, 2, \ldots, ab$ in order in a rectangular array with b rows and a columns.)

5. Prove that if $(n_1, n_2) = 1$, the congruences $x \equiv a_1(n_1)$ and $x \equiv a_2(n_2)$ have a common solution and that any two common solutions are congruent mod $n_1 n_2$.

6. If p is a prime >3, prove that

$$p(p^2-1)(p^2-4)(p^2-9) \equiv 0(2.3.5.7).$$

7. Prove that $2^{n-1}(2^n-1)$ is a *perfect number* (i.e. is equal to the sum of all its divisors excluding the number itself) provided that 2^n-1 is prime. Hence find the 3 smallest such perfect numbers.

8. Prove that the number ... *fedcba* is divisible by 7, 11 or 13 according as the number $cba-fed+ihg- \ldots$ is so divisible.

9. If the sides of a right-angled triangle are co-prime integers, show that they are of the form m^2+n^2, m^2-n^2, $2mn$ where m, n are integers.

10. Show that if an integer of the form $4n+3$ is expressed as a product of integers, then one at least of these integers is also of the form $4n+3$. Show that each pair of integers x_i, $x_j (i \neq j)$ chosen from the sequence $x_1, x_2, \ldots$ defined by

$$x_1 = 1, \quad x_{n+1} = 4x_1x_2 \ldots x_n+3 \quad (n \geqslant 1)$$

are co-prime.

Deduce that there are an infinity of prime numbers of the form $4n+3$.

(Cambridge Open Scholarship.)

11. Let N_+, N_- be the number of positive integers of the form $3k+1$, $3k-1$, respectively, with integral k, which divide a given positive integer n, both 1 and n being counted. Show that $N_+ \geqslant N_-$.

(Cambridge Open Scholarship.)

12. Prove that if a square ends with a 6 the figure in the tens' place is odd, otherwise the figure in the tens' place is even.

13. A sequence of integers u_n is generated by the recurrence relation $u_{n+1} = u_n + u_{n-1}$. Show that the sequence of remainders when the u_n are divided by a fixed integer k is periodic. Deduce that if $u_0 = 1$ and $u_1 = 1$ then some u_n is divisible by k. By considering the case $k = 5$, show that this last result is not true for all pairs of initial values u_0 and u_1.

(Cambridge Open Scholarship.)

7

POLYNOMIALS

7.1. Definition and algebra of polynomials

The sets which we consider in this chapter are those of polynomials in a variable x. A typical polynomial is $a_n x^n + a_{n-1}x^{n-1} + \ldots + a_1 x + a_0$ where the numbers $a_n, a_{n-1}, \ldots, a_0$ are called *coefficients* and, if $a_n \neq 0$, n is the *degree* of the polynomial. Most of the work will be familiar to the student from elementary algebra and much may seem obvious: its importance again lies in the fact that we are applying ordinary algebraic rules to sets other than those of numbers, our elements here being polynomials. Of particular importance in this chapter is the fact that we will discover our algebra of polynomials to be very similar to that of the integers and many of the processes of chapter 4 can therefore be applied.

The coefficients of a polynomial may be elements from a wide variety of sets: for the moment we will think of them as real numbers.

The 'x' in the polynomial $P(x) \equiv a_n x^n + a_{n-1}x^{n-1} + \ldots + a_0$ may be thought of in two ways. In elementary work we usually take it to stand for some number, its insertion in $P(x)$ giving P a numerical value, thus making $P(x)$ a 'polynomial function' of x. If two polynomials have the same value for some value x, we say that $P(x) = Q(x)$, using the *equals* sign. As we progress in algebra we begin to think of the polynomial as an element in its own right, with x becoming merely a symbol whose value is immaterial. Two polynomials are the same only if they have identical coefficients, we write $P(x) \equiv Q(x)$ with the *identity* sign, and $P(x)$, $Q(x)$ have the same value if we give x *any* numerical value.

Since we are interested chiefly in a polynomial as an element of a set the second of the two approaches above will be the one which we adopt in most of our work. The 'x', a symbol which could be replaced by any other symbol or even omitted in certain circumstances, is called an *indeterminate*. Polynomials will be denoted by capital letters, and the indeterminate x will often be omitted, thus our polynomial elements will be written as $P, Q, \ldots$. If P and Q are identical we will say that $P = Q$, using the equals sign instead of the identity sign for convenience. This need cause no confusion, provided we interpret $P = Q$ to mean 'P is the same polynomial as Q', *not* 'P and Q have the same value'.

We may add, subtract, or multiply polynomials in the usual way. Thus if

$$P \equiv a_0 + a_1 x + \ldots + a_n x^n,$$

$$Q \equiv b_0 + b_1 x + \ldots + b_m x^m, \quad \text{where} \quad m < n,$$

we have

$$P \pm Q \equiv (a_0 \pm b_0) + (a_1 \pm b_1)x + \ldots + (a_m \pm b_m)x^m + a_{m+1}x^{m+1} + \ldots + a_n x^n,$$

$$PQ \equiv a_0 b_0 + (a_0 b_1 + a_1 b_0)x + (a_0 b_2 + a_1 b_1 + a_2 b_0)x^2 + \ldots + a_n b_m x^{m+n}.$$

Abstract definition of polynomials

Rather than having to depend on our background knowledge of elementary algebra, it is useful to have a completely abstract definition of polynomials in terms of coefficients, and to define the processes of algebra in terms of this definition. We may then prove the basic laws of algebra direct from the definition, avoiding any intuitive notions of what should be true. When we frame the definition we have, of course, in the back of our minds the elementary work, but the abstractness makes the work theoretically independent of this.

We define a polynomial P as the ordered set of coefficients $(a_0, a_1, \ldots)$, where every a_i after a certain one is zero. The a_i's may belong to a variety of sets, but for the moment they are real numbers. It is more convenient to take the infinite set of coefficients with all except a finite number of them being zero. If the last non-zero a_i is a_n then n is the *degree* of the polynomial, and is written $d(P)$. (By convention we say that

$$d(0, 0, 0, \ldots) = -\infty.)$$

If $P = (a_i)$ and $Q = (b_i)$, we define $P+Q$ to be (a_i+b_i) and $P-Q$ to be (a_i-b_i).

PQ is defined as (c_i) where

$$c_i = \sum_{r+s=i} a_r b_s. \tag{1}$$

See p. 124 for a note on the derivation of the usual notation.

Algebra of polynomials

We now investigate the applicability or otherwise of the ordinary fundamental laws of algebra as applied to polynomials. In a rigorous treatment we would prove these from the abstract definition above, but here we will merely indicate the results, which are all fairly immediate, either abstractly or using intuitive ideas.

Addition is commutative and associative.

Subtraction. The polynomial (0, 0, ...) does not alter any other when added to it. We call this the *zero* polynomial. The *negative* of $P = (a_i)$ is $-P = (-a_i)$ and has the property that when added to P it gives the zero. $P-Q = P+(-Q)$. The Cancellation Law holds: $P+Q = P+R \Rightarrow Q = R$.

Multiplication. This again is commutative and associative, and the Distributive Laws hold. The polynomial (1, 0, 0, ...) has the property that when multiplied by P it does not alter P, and is called the *unity*.

Theorem 7.1.1. *Polynomials have no zero divisors, i.e.* $PQ = 0 \Rightarrow$ *either* $P = 0$ *or* $Q = 0$.

Suppose P and Q are not zero. Then if $P = (a_i)$, $Q = (b_i)$ and $d(P) = n$, $d(Q) = m$, a_n and b_m are not zero. But $PQ = (c_i)$ where $c_{m+n} = a_n b_m \neq 0$, and so $PQ \neq 0$.

Corollary. $d(PQ) = n+m$.

Theorem 7.1.2. *The Cancellation Law of multiplication is true for polynomials, i.e.* $PQ = PR \Rightarrow Q = R$ *provided* $P \neq 0$.

For $PQ-PR = 0$. $P(Q-R) = 0$ and so by theorem 7.1.1 either $P = 0$ or $Q-R = 0$. Hence if $P \neq 0$, $Q = R$.

Division. This is not always possible, i.e. $PX = Q$ does not always have a solution, though it may have in certain cases.

There can never be more than one solution by the Cancellation Law, unless $P = Q = 0$.

Example. If $P \equiv 1+x$ and $Q \equiv 1+x^2$ there is no solution, while if $Q \equiv 1-x^2$ there is the unique solution $X \equiv 1-x$.

Order and induction. There is no simple way of ordering the set of polynomials, and direct induction is not possible. We may use a modified induction on the degree, i.e. we assume a result true for all polynomials of degree k and then deduce it for all those of degree $k+1$.

To sum up we see that polynomials may be added, subtracted, and multiplied, there is a unity, and there are no zero divisors. Division is not always possible. But these are the basic properties of the set of integers (except that the latter are ordered) and thus we would expect the set of polynomials to have many properties in common with the integers. This is so, and much of this chapter will be analogous to chapter 4.The reader should carefully study the similarities and differences, and note how these depend on the underlying basic laws. Both sets form what is called an 'Integral Domain'.

7.2. The coefficients of a polynomial

Although in the last section we took the coefficients a_i to be real numbers, this is by no means necessary. For the abstract definition given to allow addition, subtraction and multiplication it is sufficient that the coefficients be elements of a set within which these processes are possible. Possible sets are the integers, the rationals, real numbers, complex numbers, or residue classes modulo any integer. The proof of theorem 7.1.1 depended on the fact that $a_n \neq 0$ and $b_m \neq 0 \Rightarrow a_n b_m \neq 0$, i.e. that the set in which the coefficients lie has itself no zero divisors. Thus if the set of coefficients has no zero divisors then neither has the set of polynomials over it. But even if the set of coefficients admits of division (e.g. the real numbers) the set of polynomials will not always do so. Furthermore the set of polynomials will always be infinite, even when the coefficients form a finite set.

We will see that it is important to specify the set to which the coefficients are required to belong. If the coefficient set is A, we talk of polynomials 'over the set A'.

Note that the set of polynomials always contains a subset which behaves like the set of coefficients, viz. all those polynomials of degree 0.

7.3. Divisibility and irreducibility

One of the most important subjects in our study of the integers was that of divisibility and primes. The same is true of polynomials, and the definitions are similar.

Q is a *factor* of P if there exists a polynomial R such that $P = QR$. We say that P is *divisible* by Q or is a *multiple* of Q. This means of course that we can divide P by Q. (Neither P nor Q is zero.) We write Q/P.

A factor of the unity 1 is called a *unit* (distinguish carefully between unit and unity). For the set of integers the only units were 1 and -1, but for polynomials there may be more.

Theorem 7.3.1. *If the coefficients have no zero divisors so that* $d(PQ) = d(P)+d(Q)$

(i) *A unit must have degree* 0.

(ii) *The units correspond to the units in the set of coefficients. Thus the polynomial* U *(i.e.* $(u, 0, \ldots)$*) is a unit if and only if the number* u *is a unit in the coefficient set.*

(i) If U is a unit there exists R such that $UR = 1$. Therefore $d(U)+d(R) = d(UR) = 0$ and $d(R) \geqslant 0$. Hence $d(U) = 0$.

(ii) If $(u, 0, \ldots)$ is a unit, R must also be a unit and so have degree 0. Let $R = (r, 0, \ldots)$. Then $1 = ur$ and so u is a unit of the coefficient set. Conversely if u is a unit of the set of coefficients, there exists r such that $ur = 1$. Hence $U = (u, 0, \ldots)$ is a unit of the set of polynomials, since if $R = (r, 0, \ldots)$ $UR = 1$.

By the above theorem we see that in the set of polynomials over the real numbers any polynomial of degree 0, i.e. any constant, is a unit (except 0 itself, which is always excluded). Similarly, for polynomials over the complex numbers, and those over the rationals. For polynomials over the integers the only units are 1 and -1.

The importance of units is shown by the following theorems.

Theorem 7.3.2. *If* Q *is a factor of* P *and* U *is a unit, then* UQ *is a factor of* P.

There exists U' such that $UU' = 1$. There exists R such that $P = QR$. Hence $P = UU'QR = (UQ)(U'R)$ and so UQ is a factor of P.

Theorem 7.3.3. *If U is a unit it is a factor of all polynomials P.* For if $UU' = 1$, $P = U(U'P)$ and U is a factor of P.

We will shortly be dealing with primes and prime factorisation for polynomials. We see from the above theorems that we must ignore the units as possible factors, and that uniqueness of factorisation must also take no account of possible unit factors. We will work, so to speak, *modulo units*. (Just as in residues we work *modulo n*, ignoring multiples of n.) The same technique was necessary with integers, though rather obscured by the scarcity of units. The only unit integers are 1 and -1, and ignoring these means in effect that we always deal with *positive* primes when factorising, merely putting a '$-$' in front when factorising a negative integer.

Examples. $x+1$ is a prime polynomial, having no factors in the usual sense of the term. But of course any real number λ is a factor, since $x+1 = \lambda[(1/\lambda)x+(1/\lambda)]$. However, λ is a unit and so $x+1$ is still prime when we work modulo units.

x^2-1 has the unique factors $x+1$ and $x-1$, but these are only unique if we ignore multiplication by units, since $\lambda(x+1)$ and $(1/\lambda)(x-1)$ are also a pair of factors.

Associates. If U is a unit, the polynomials P and UP are called associates, and factorisation is unaltered if we replace a factor by any associate (possibly with adjustment to other factors).

Theorem 7.3.4. *The relation of being associates is an equivalence relation.*

Reflexive. $P = 1.P$ and 1 is a unit.

Symmetric. If $P = UQ$ and U' is such that $UU' = 1$, $U'P = UU'Q = Q$.

Transitive. If $P = UQ$ and $Q = VR$ then $P = UVR$. But UV is a unit. For if $UU' = VV' = 1$,

$$UU'VV' = (UV)(U'V') = 1.$$

Hence our set of polynomials divides into equivalence classes, any two members of the same class being associates. If Q is a factor of P then so is any member of the class containing Q. (Of course Q is a factor of any member of the class containing P since Q must divide UP whatever U is, and therefore particularly when U is a unit.)

For polynomials over the reals, rationals or complex numbers the members of the class containing P are merely $\{\lambda P\}$ for any non-zero λ in the set of coefficients.

Irreducibility

A polynomial which has the property analogous to a prime number is called irreducible. Any polynomial has as divisors any unit and any associate. One which has no divisors except these is called *irreducible.*

Note that it is important to specify the set to which the coefficients are deemed to belong. Thus $x^2 - 2$ is irreducible over rational coefficients, but is *not* irreducible over real coefficients, since in the latter case it may be factorised as $(x+\sqrt{2})(x-\sqrt{2})$. Similarly, x^2+1 is irreducible over both the rationals and reals, but not over the set of complex numbers.

The associates and units are sometimes known as trivial factors, any others are *non-trivial*; thus a polynomial is irreducible if it has no non-trivial factors. In the case where the set of coefficients admits of division (e.g. rationals, or reals, or complex numbers, etc., but not the integers) any non-trivial factor must be of degree greater than 0 and less than the degree of the polynomial, since any factor of degree 0 is a unit. This need not be so if the coefficient set does not admit of division; thus $x+1$ is a non-trivial factor of $3x+3$ over the integers, as is the number 3 itself, since neither of these is a unit or associate in this set.

***Theorem* 7.3.5.** *The irreducible polynomials over the complex numbers are all those of the first degree.*

By the fundamental theorem of algebra any polynomial of degree greater than one can be factorised. Those of degree 0 are units, while the only factors of those of degree 1 are either units or associates.

***Theorem* 7.3.6.** *In the set of polynomials over the reals, all those of degree* 1 *are irreducible, while the only other irreducible ones are of degree* 2.

Any factor of a first degree polynomial is a unit or associate. Since complex roots of a polynomial equation with real coefficients occur in conjugate pairs, any such polynomial may be factorised into the product of linear and quadratic factors. Hence the only irreducible ones have degree 1 or 2. Of course not all of those of degree 2 are irreducible—in fact ax^2+bx+c is irreducible if and only if $b^2 < 4ac$.

There is no easy method of determining whether or not a polynomial over the rationals or integers is irreducible. We will investigate this question a little further in §7.8.

Theorem 7.3.7. *If Q is a factor of P and P is a factor of Q then P and Q are associates.*

(This is analogous to theorem 4.4.1.)

$P = QR$ and $Q = PS$ for some R and S. Hence $P = PRS$ and so by the Cancellation Law $RS = 1$. Thus R and S are units and so P and Q are associates.

7.4. Highest common factor

As we did with the integers we continue our study of polynomials by investigating the H.C.F. of two or more, leading to the consideration of prime factorisation. The procedure will be very similar to that of chapter 4.

The remark made at the beginning of §4.5 about the need for a fairly elaborate approach applies even more here. The reader may think that the H.C.F. is best found by factorising each polynomial and selecting all common factors, and of course in many particular examples this is the easiest way. But apart from the fact that uniqueness of factorisation into primes has not yet been proved it is not always easy to find the factorisation by *ad hoc* methods. In the case of polynomials over the complex or real numbers we know that in theory we can always factorise into irreducible factors of degree 1 or 2, but for polynomials of high degree (5 or above) there is no general method of finding the factors. For polynomials over the rationals the position is even worse, for it is not always easy to discover whether or not a given factor is irreducible, and so we have no guarantee that a factorisation which we may have found is the final one. Our method avoids the problem of actually having to factorise the given polynomials in order to find their H.C.F. Later we will prove that prime factorisation is unique, but we will still have no general method of finding the actual factors.

The work of this section and the next depends on the possibility of division within the set of coefficients. A set in which addition, subtraction, multiplication and division are all possible according to the usual rules is called a *field*, and our work holds for any field. Thus it applies to coefficients in the field of rationals, reals or complex numbers, but not to those in the set of integers. It also applies to coefficients in the set of residues modulo a prime, but not modulo a composite number. The reader may like to think of the coefficients as being reals, for simplicity, but he should note that any properties of the coefficient set that he uses apply to the other fields also.

Definition. *The polynomial H is an* H.C.F. *of the two polynomials A and B (non-zero) if*

(i) *$H|A$ and $H|B$, i.e. H is a factor of both A and B;*

(ii) *if $C|A$ and $C|B$ then $C|H$, i.e. any common factor of A and B is a factor of H.*

Theorem 7.4.1. *The* H.C.F. *is unique to within a class of associates.*

If H and H' are both H.C.F.'s of A and B, they are both common factors by (i) and so by (ii) we have $H|H'$ and $H'|H$. The result follows by theorem 7.3.7.

The existence of an H.C.F. is shown by the Euclidean algorithm as before.

Theorem 7.4.2. *The division algorithm for polynomials.*

Given any two polynomials A and B, $B \neq 0$, then there exist unique polynomials Q and R such that $A = BQ+R$ and such that $d(R) < d(B)$. (R may be 0.)

This gives a formal statement of the usual process of division to give a quotient Q and remainder R.

To prove that Q and R exist we use induction, and assume the result true for all polynomials of degree less than A, i.e. if A' has degree less than $d(A)$, then Q' and R' exist, $d(R') < d(B)$ such that $A' = Q'B+R'$.

If $d(A) < d(B)$ the result is trivial, with $Q = 0$. Suppose then that $d(A) \geqslant d(B)$. Let $A \equiv a_n x^n + \ldots$ and $B \equiv b_m x^m + \ldots$, and consider the polynomial $A' \equiv A-(a_n/b_m)x^{n-m}B$. Then A' has degree at most $(n-1)$, i.e. less than $d(A)$, and so by the inductive hypothesis we can find Q' and R', $d(R') < d(B)$, such that $A' = Q'B+R'$. Hence $A = QB+R$ where

$$Q \equiv Q'+(a_n/b_m)x^{n-m} \quad \text{and} \quad R = R',$$

so that $d(R) < d(B)$ as required. The result is true for $d(A) < d(B)$ and so the induction can start. Hence it is true generally.

To prove uniqueness, suppose that $A = BQ+R$ and $A = BQ'+R'$ with $d(R)$ and $d(R')$ both $< d(B)$. Then

$$B(Q-Q') = R-R'.$$

But the degree of the left-hand side $= d(B)+d(Q-Q') \geqslant d(B)$ and $d(R-R') < d(B)$. Hence both sides are 0 and so $Q = Q'$ and $R = R'$.

It is in the above proof that we need the fact that the set of coefficients must be capable of division, since in forming A' we need to form the ratio a_n/b_m.

To find Q and R explicitly we use the familiar division process, which in effect finds A', repeats to find A'' and so on.

Euclid's algorithm for polynomials. *To find an* H.C.F. *of A and B.*

By the division algorithm we can find Q_1 and R_1 such that

$$A = BQ_1+R_1 \quad \text{with} \quad d(R_1) < d(B).$$

Similarly, we complete the series of equations

$$\begin{aligned} A &= BQ_1+R_1, \\ B &= R_1Q_2+R_2, \\ R_1 &= R_2Q_3+R_3, \\ &\ldots\ldots\ldots\ldots\ldots\ldots \\ R_{n-2} &= R_{n-1}Q_n+R_n \\ R_{n-1} &= R_n Q_{n+1}, \end{aligned}$$

where $d(B) > d(R_1) > d(R_2) > \ldots$ and so the series must terminate after at most $d(B)+1$ steps.

Then R_n is an H.C.F. of A and B, the proof being exactly as in §4.5.

Hence every two polynomials have an H.C.F. unique within associates. We call the polynomial within this class of associates which has leading coefficient unity *the* H.C.F., and write it as (A, B). If the H.C.F. is 1 we say that A and B are co-prime.

Theorem 7.4.3. *If $H = (A, B)$ then there exist polynomials S and T such that $H = SA+TB$.*

Proved as in theorem 4.5.3.

Corollary. *If A and B are co-prime there exist S and T such that $1 = SA+TB$.*

The H.C.F. of more than two polynomials may be defined and found as in §4.5.

7.5. Prime factorisation of polynomials

We will prove here that any polynomial with coefficients in a field may be expressed as the product of polynomials which are irreducible over that field, and that this may be done in one and only one way. For polynomials over the complex field all factors will be linear, for those over the reals they will be linear or quadratic, while for coefficients in the rational field they may be of various degrees. Note that our theorems do *not* give us a method of actually finding the factors. The uniqueness of factorisation will work only modulo units and associates.

Theorem 7.5.1. *If P is irreducible and $P|AB$ then either $P|A$ or $P|B$.*

Proved as in theorem 4.6.1.

Theorem 7.5.2. *The unique factorisation theorem.*

Any polynomial with coefficients in a field may be expressed in the form $cP_1P_2 \ldots P_n$, where $P_1, P_2, \ldots, P_n$ are irreducible polynomials and c is a constant, and this expression is unique except for the order of the factors, modulo associates.

We prove this as in theorem 4.6.2, by using 7.5.1 above, using induction on the degree of the polynomial in the possibility part of the proof, and in the uniqueness part noting that any factor may be replaced by an associate with a corresponding adjustment to the unit c.

7.6. Lowest common multiple

The work is completely analogous to §4.7 with obvious modifications to take account of associates.

7.7. Zeros of a polynomial: the remainder theorem

For a polynomial $P(x)$ with coefficients in any set that admits of addition, subtraction and multiplication we may replace x by any number element c in the coefficient set. Thus the polynomial becomes a function over this set, and we write its value (which will itself be an element of the set) as $P(c)$.

Hence if

$$P(x) \equiv a_0 + a_1 x + \ldots + a_n x^n,$$

$$P(c) = a_0 + a_1 c + \ldots + a_n c^n.$$

An element c such that $P(c) = 0$ is called a *zero* of P.

The remainder of this section applies only to polynomials over a field, since it depends on the division algorithm. Note that a field cannot have zero divisors, since if $ab = 0$ and $a \neq 0$, we know that $1/a$ exists and so $(1/a)ab = 0$, i.e. $b = 0$.

Theorem 7.7.1. *The remainder theorem.*

If $A(x)$ is divided by $(x-c)$ the remainder is a constant and is equal to $A(c)$.

Apply the division algorithm with $B = x - c$. Then

$$d(R) < d(B) = 1$$

and so R is a constant. Then $A(x) \equiv (x-c)Q(x) + R$ and this is of course an identity. Hence, putting $x = c$ on both sides we have $A(c) = R$.

Corollary. *The factor theorem.*

$(x-c)$ is a factor of $A(x)$ if and only if $A(c) = 0$, i.e. if and only if c is a zero of $A(x)$.

Theorem 7.7.2. *A non-zero polynomial $P(x)$ of degree n has at most n distinct zeros.*

(Compare §4.3, example 2 and theorem 6.5.1.)

The proof is by induction. We assume that no polynomial of degree $n-1$ has more than $n-1$ distinct zeros. Let $c_1, c_2, \ldots, c_m$

be distinct zeros of $P(x)$. Then by the factor theorem $P(x)$ has a factor $(x-c_1)$ and so $P(x) \equiv (x-c_1)Q(x)$ where $Q(x)$ is of degree $n-1$. If $i \neq 1$, c_i is a zero of $P(x)$ and so $P(c_i) = 0$. Hence $(c_1-c_i)Q(c_i) = 0$ and, as the coefficients have no zero divisors, it follows since $c_i \neq c_1$ that $Q(c_i) = 0$ and c_i is a zero of $Q(x)$. Hence $c_2, c_3 \ldots c_m$ are distinct zeros of $Q(x)$ and by the inductive hypothesis $Q(x)$ has at most $n-1$ such. Hence $m \leqslant n$ and the result is true for $P(x)$.

But a polynomial of degree 0, being a constant, has no zeros unless it is itself zero, and so the result follows by induction.

Corollary 1. *If $P(x)$ is a polynomial of degree n which vanishes for more than n distinct values of x, then $P(x)$ is the zero polynomial and vanishes for* all *values of x.*

Corollary 2. *If $c_1, c_2, \ldots, c_m$, where $m \leqslant n$, are distinct zeros of $P(x)$, $P(x)$ has $(x-c_1)(x-c_2) \ldots (x-c_m)$ as a factor.*

For $c_2, c_3, \ldots, c_m$ are zeros of $Q(x)$ in the proof of the theorem and the result follows by induction.

7.8. Gauss's theorem

We mentioned in §7.3 that there is no easy method of determining whether or not a given polynomial is irreducible over the rationals or integers, and we can in fact have such irreducible polynomials of any degree. Thus $x^{2n}+1$ is irreducible over both the rationals and the integers for any positive integer n, as is $x^{p-1}+x^{p-2}+ \ldots +x+1$, where p is prime and >2. Gauss's theorem gives us a useful criterion which can be applied in many cases. The theorem states that if a polynomial with integral coefficients is irreducible over the integers then it is also irreducible over the rationals: in other words, if such a polynomial has a factor with rational coefficients then it must have one (in fact a constant multiple of the other) with integral coefficients, and the quotient also has integral coefficients. Thus we need search only for factors over the integers.

Much of the difficulty in proving Gauss's theorem arises from the fact that although there is always a factor over the integers corresponding to one over the rationals, it may not be the same but merely a constant multiple of the latter. For example,

$x^2-1 \equiv (\frac{1}{2}x+\frac{1}{2})(2x-2)$ and has a factor $\frac{1}{2}x+\frac{1}{2}$ over the rationals. But this is *not* a factor over the integers, the corresponding such factorisation being $(x+1)(x-1)$. This leads us to adopt the definition below.

Definition. *A polynomial (over the integers or rationals) is* primitive *if its coefficients are integers and have no common factor.*

Lemma 1. *The product of two primitive polynomials is itself primitive.*

Suppose
$$P(x) \equiv a_n x^n + a_{n-1}x^{n-1} + \dots + a_0,$$
$$Q(x) \equiv b_m x^m + b_{m-1}x^{m-1} + \dots + b_0$$

are both primitive, i.e. all the a_i's and b_j's are integers and each set has no common factor. Let

$$PQ = R(x) \equiv c_{m+n}x^{m+n} + \dots + c_0.$$

It is at once obvious that all the c_k's are integers. Suppose they have a common factor p, where p is a prime. Then p is *not* a common factor of all the a_i's, or of all the b_j's, since P and Q are primitive. Let a_α and b_β be the first of the a_i's and b_j's (starting from the highest coefficients) which are not divisible by p. Then

$$c_{\alpha+\beta} = a_\alpha b_\beta + a_{\alpha+\beta}b_0 + \dots + a_{\alpha+1}b_{\beta-1} + a_{\alpha-1}b_{\beta+1} + \dots + a_0 b_{\alpha+\beta}$$

and all terms on the *RHS* except the first are divisible by p, as is $c_{\alpha+\beta}$. Hence we have a contradiction ($a_\alpha b_\beta$ is not divisible by p since neither a_α nor b_β is) and so the c_k's cannot have a common factor. Thus $R = PQ$ is primitive.

Lemma 2. *Any polynomial $P(x)$ over the rationals can be expressed in the form $cP^*(x)$ where P^* is primitive and c is a rational constant, and such an expression is unique to within + or −.*

If $P(x) \equiv a_n x^n + \dots + a_0$ let h be the L.C.M. of the denominators of $a_n, \dots, a_0$. Then $P = (1/h)P'$ where P' has integral coefficients. Let k be the H.C.F. of the coefficients of P'. Then $P = (k/h)P^*$ where P^* is primitive. Put $c = k/h$.

To prove uniqueness let $P = cP^* = c'P^{*\prime}$. Then

$$P^* = (c'/c)P^{*\prime} = (u/v)P^{*\prime}$$

say, where u and v are co-prime integers. Thus $vP^* = uP^{*\prime}$ and so, since $P^{*\prime}$ is primitive v must be a factor of u. Similarly, u is a factor of v, since P^* is primitive, and so $u = \pm v$. Hence $c = \pm c'$ and so $P^* = \pm P^{*\prime}$.

Theorem 7.8.1. *Gauss's theorem.*

If $P(x)$ is a polynomial with integral coefficients which is irreducible over the integers, then it is irreducible over the rationals.

Suppose P can be factorised over the rationals, $P = QR$ say. By lemma 2 we can write

$$P = cP^*, \quad Q = dQ^*, \quad R = eR^*,$$

where P^*, Q^*, R^* are primitive, c is an integer (since P is over the integers), and d and e are rational constants. Then

$$cP^* = deQ^*R^*$$

and by lemma 1 Q^*R^* is primitive. Hence by the uniqueness part of lemma 2 we must have $P^* = \pm Q^*R^*$ and $c = \pm de$, giving $P = \pm cQ^*R^*$. But since c is an integer and Q^*, R^*, being primitive, are over the integers, this gives a factorisation over the integers. (c may be taken with either Q^* or R^* if we wish, to give 2 integral factors only.)

Example. Let $P \equiv 3x^2-3$, $Q \equiv \frac{1}{2}x+\frac{1}{2}$, $R \equiv 6x-6$.

Then $c = 3, d = \frac{1}{2}, e = 6, P^* \equiv x^2-1, Q^* \equiv x+1, R^* \equiv x-1$. and of course $c = de$ and $P^* = Q^*R^*$, so that we get

$$P \equiv 3(x+1)(x-1)$$

as the factorisation over the integers. This may be written either as $(3x+3)(x-1)$ or $(x+1)(3x-3)$.

Note on the importance of lemma 1. We may think that lemma 1 hardly enters into the proof. It is, however, fundamental to it, for if it were not true then Q^*R^* might have a constant factor w say, where w is an integer. Then de need not be an integer (it could be a fraction with denominator w, still making the coefficients of P integral) and the factorisation deQ^*R^* would not be over the integers.

Theorem 7.8.2. *Working over the integers, if*

$$P \equiv a_n x^n + \ldots + a_0$$

has a factor $b_m x^m + \ldots + b_0$, *then* b_m *is a factor of* a_n *and* b_0 *is a factor of* a_0.

Suppose the remaining factor is $c_{n-m}x^{n-m} + \ldots + c_0$. Then $a_n = b_m c_{n-m}$ and $a_0 = b_0 c_0$ and the result follows.

This theorem narrows our search for factors over the integers considerably, though the problem may still be a difficult one if the factors are of high degree. The problem of finding linear factors is now simple, and hence so is that of finding rational roots, by the factor theorem. For any rational root of $P(x) = 0$ must be of the form r/s where r is a factor of a_0 and s is a factor of a_n. [By Gauss's theorem the factor theorem, and so also theorem 7.7.2, is true over the integers. For if $x-c$ is a factor over the rationals and c is an integer then $x-c$, being primitive, is also a factor over the integers.]

By Gauss's theorem we may reduce the problem of factorising over the rationals to one of factorising over the integers. For suppose P is a polynomial with *rational* coefficients. Then if h is the L.C.M. of the denominators of the coefficients, $P = (1/h)P'$, where P' is over the integers. By Gauss's theorem any rational factor of P' corresponds to an integral factor.

Thus to find a rational root of a polynomial equation over the rationals we first clear of fractions by multiplying the equation by the L.C.M. of the denominators of the coefficients and then proceed as before, testing all factors $(sx-r)$ where r and s are factors of the lowest and highest coefficients, respectively.

Example. The equation $x^{p-1} + x^{p-2} + \ldots + x + 1 = 0$ when p is prime and >2 has no rational root. For such a root must be 1 or -1 and we can easily verify that neither of these is in fact a root.

We have not of course proved that the polynomial is irreducible, since it may have factors of degree higher than 1. This particular polynomial is in fact irreducible.

Theorem 7.8.3. *Any rational root of a polynomial with integral coefficients whose leading coefficient is* 1 *is an integer, which must be a factor of the constant term.*

This is almost immediate, since by theorem 7.8.2 and Gauss's theorem any linear factor of such a polynomial must be of the form $\pm(x \pm r)$ where r is a factor of a_0. The following direct proof may be of interest.

Suppose r/s is a rational root, in its lowest terms, of

$$x^n + a_{n-1}x^{n-1} + \dots + a_0 = 0.$$

Then $(r/s)^n + \dots + a_0 = 0$ and so

$$r^n + a_{n-1}r^{n-1}s + \dots + a_0 s^n = 0.$$

Thus s is a factor of r^n which is impossible since r and s are co-prime. Therefore $s = 1$ and r is a factor of a_0.

7.9. Various sets of coefficients

Throughout this chapter we have had to specify the sets to which the coefficients may belong in order for our theorems to apply. If the coefficients belong to a field, such as the rationals, reals or complex numbers, or even the finite field of residues modulo a prime number, then all our results are valid. This is not true for other sets of coefficients, such as the integers or residues modulo a composite number. In this section we will summarise the important conditions that we must place on our coefficient set.

Cancellation Law

This applies to polynomials provided it does so for the set of coefficients. Thus it holds for all our basic sets except that of residues modulo a composite number. For example in the set of residues modulo 4 the product $(2x+2)(2x-2)$ is zero while neither factor vanishes.

H.C.F.

Our work on this is valid over any field, but not over other sets. Thus the division algorithm is not true for polynomials over the integers, e.g. we cannot divide x^2 by $2x+1$ to obtain a remainder of degree 0. Neither is theorem 7.4.3 true: the H.C.F. of 2 and x is 1 but no polynomials S and T exist such that $S.2 + T.x \equiv 1$, for the constant term of S would have to be $\frac{1}{2}$, which is not an integer.

For polynomials over the set of residues modulo 6, say, the division algorithm does not apply to x^2 and $2x$ (for 2 does not divide into 1). Not only does theorem 7.4.3 not hold, but there is not even an H.C.F. in this case, as we will see in the next paragraph.

Prime factorisation

The work of §7.5, which depends on that of §7.4, is valid only for coefficients over a field. The *result* of theorem 7.5.2, that factorisation into irreducible polynomials is unique modulo units, is however true in certain other cases, the most important being those where the set of coefficients has the same property. Thus since unique factorisation holds for the set of integers it is also true for polynomials over the integers. We do not prove this theorem here, but a proof may be found in any standard work on integral domains (see, for example, Birkhoff and MacLane, *A Survey of Modern Algebra*, chapter III, theorem 16). Note that this property of unique factorisation guarantees the existence and uniqueness of H.C.F. and L.C.M., by the elementary method of expressing both the polynomials in terms of their prime factors and picking out the relevant factors. But theorem 7.4.3 does not hold, as we have already seen.

Polynomials over residues with a composite modulus do not have unique factorisation. Thus modulo 6,

$$x^2+x \equiv x(x+1) \equiv (x+3)(x-2) \text{ and } 2x \equiv 2.x \equiv 2(x+3).$$

Neither is there an H.C.F., for x and $x+3$ are both common factors of the two polynomials given, but $x(x+3)$ is not. The absence of the Cancellation Law makes any discussion of primes or factors rather pointless in this case (e.g. while for the set of residues modulo 6 the only units are 1 and 5, there are no primes, since $2 = 2.4$, $3 = 3.3$, $4 = 4.4$).

Zeros

Our proof of the remainder theorem, which depends on the division algorithm, is applicable only when the set of coefficients is a field, and the proof of the factor theorem given in §7.7 depends on the remainder theorem. These theorems are, however, true generally (see exercises 7B, no. **10**) and a direct proof of the factor theorem follows.

Theorem 7.9.1. *The factor theorem for any set of coefficients.*

If $A(c) = 0$,

$$A(x) \equiv A(x) - A(c) \equiv a_n(x^n - c^n) + \ldots + a_1(x-c). \quad (1)$$

But for any m,

$$x^m - c^m \equiv (x-c)(x^{m-1} + x^{m-2}c + \ldots + c^{m-1})$$

by ordinary expansion, and this is valid for any coefficient set which admits of addition, subtraction and multiplication. Hence $A(x)$ has a factor $x-c$ from (1). Conversely if $A(x) \equiv (x-c)B(x)$ then clearly $A(c) = 0$.

Thus theorem 7.7.2, which used the factor theorem, is now established for any coefficient set with no zero divisors. It is therefore true for polynomials over the integers, but not for those over residues modulo 8 (say). Thus the polynomial $x^2 - 1$ has four zeros modulo 8, viz. 1, 3, 5, 7.

Note that if the coefficients form a finite set the polynomial may be zero for *all* values of x but not be the zero polynomial. For residues modulo n such a polynomial is given by

$$x(x-1)(x-2) \ldots (x-n+1),$$

which cannot be the zero polynomial since the leading term, x^n, is not zero. But corollary 1 of theorem 7.7.2 still holds if the modulus is prime (in the case given above the polynomial is of degree n and vanishes for n values only). Of course this corollary is not necessarily true for a composite modulus. Thus modulo 8 the polynomial $(x^2-1)(x^2-4)x^2$, of the sixth degree, vanishes for 8 distinct values of x (in fact for all values of x) and yet is not the zero polynomial.

7.10. Rational functions

We have seen that the set of polynomials behaves very much like the set of integers, in that they can both be added, subtracted, or multiplied but not divided. In §5.1, in order to introduce division in connection with the integers, we formed a new set, that of the rationals, in an abstract manner starting with the integers, and the only properties of integers that were used were the basic laws of addition, subtraction and multiplication, together with the fact of there being no zero divisors.

Hence from any set with these properties (i.e. from any *integral domain*) we may form a corresponding set of rationals. In particular we can do so from the set of polynomials over any coefficient set which is itself an integral domain (thus over any of the sets which we have been considering except the residues modulo a composite number). The fractions that we obtain are called *rational functions*. A typical one is $P(x)/Q(x)$ where $P(x)$ and $Q(x)$ are polynomials over the given set of coefficients, and $P(x)/Q(x)$, $R(x)/S(x)$ are identical if $PS = QR$.

Within the set of rational functions division is always possible, there are no primes and every element is a unit.

Note that the polynomials over the integers and those over the rationals give rise to the same set of rational functions; by multiplying the numerator and denominator by suitable integers any rational function formed by two polynomials over the rationals becomes the quotient of polynomials over the integers.

Note

To derive the usual notation we note that

$$(a, 0, 0, \ldots)+(b, 0, 0, \ldots) = (a+b, 0, 0, \ldots)$$

and $$(a, 0, 0, \ldots).(b, 0, 0, \ldots) = (ab, 0, 0, \ldots)$$

and so we identify $(a, 0, 0, \ldots)$ with the coefficient a.

Then $$a(a_i) = (a, 0, 0, \ldots)\,(a_i) = (aa_i).$$

Now write x for $(0, 1, 0, \ldots)$.

Then by applying equation (1) (p. 107),

$$x^2 = (0, 0, 1, 0, \ldots),$$

$$x^3 = (0, 0, 0, 1, 0, \ldots)$$

and so on, and hence

$$\begin{aligned}(a_0, a_1, a_2, \ldots) &= (a_0, 0, 0, \ldots)+a_1(0, 1, 0, \ldots)+a_2(0, 0, 1, \ldots)+\ldots \\ &= a_0+a_1x+a_2x^2+\ldots.\end{aligned}$$

Worked exercises

1. Use Euclid's algorithm to find the H.C.F. of $x^7-x^5-x^4+x^2$ and x^5-x and express it in the form $SA+TB$.

(To save space we will state the quotients and remainders obtained by the progressive divisions—these may be found by the usual division process, which is elementary though tedious.)

We obtain

$$x^7-x^5-x^4+x^2 = (x^5-x)(x^2-1)-x^4+x^3+x^2-x,$$
$$x^5-x = (-x^4+x^3+x^2-x)(-x-1)+2x^3-2x,$$
$$-x^4+x^3+x^2-x = (2x^3-2x)(-\tfrac{1}{2}x+\tfrac{1}{2}),$$

Hence an H.C.F. is $2x^3-2x$ and so *the* H.C.F. is x^3-x.

To express the H.C.F. in terms of the given polynomials, we have

$$\begin{aligned}2x^3-2x &= x^5-x-(-x^4+x^3+x^2-x)(-x-1)\\ &= x^5-x-(-x-1)(x^7-x^5-x^4+x^2-(x^5-x)(x^2-1))\\ &= (x+1)(x^7-x^5-x^4+x^2)+(-x^3-x^2+x+2)(x^5-x).\end{aligned}$$

Hence

$$x^3-x = (\tfrac{1}{2}x+\tfrac{1}{2})(x^7-x^5-x^4+x^2)+(-\tfrac{1}{2}x^3-\tfrac{1}{2}x^2+\tfrac{1}{2}x+1)(x^5-x).$$

Note that this is a case where we cannot express the H.C.F. in the required form *over the integers*.

2. Factorise $A(x) \equiv 2x^5-3x^4-2x^3+3x^2-4x+6$ into irreducible factors over: (*a*) the complex numbers, (*b*) the real numbers, (*c*) the rationals, and (*d*) the integers.

There is no general method of factorising a polynomial of degree higher than 4 (there is no general method of solving a quintic), but we trust that the given polynomial will have sufficient rational zeros to make the work reasonably straightforward.

By Gauss's theorem any rational factor must be integral, and of the form $ax+b$ where a is a factor of 2 and b a factor of 6. Thus the only possible rational zeros are ± 1, ± 2, ± 3, ± 6, $\pm\tfrac{1}{2}$, $\pm 1\tfrac{1}{2}$. By testing these in turn we easily find that $+1\tfrac{1}{2}$ is the only such zero, and so by the factor theorem $(2x-3)$ is a factor. (This may have been seen by inspection in this case.)

$$A(x) \equiv (2x-3)(x^4-x^2-2) \equiv (2x-3)(x^2-2)(x^2+1) \quad (d).$$

For factors over the rationals we have

$$(2x-3)(x^2-2)(x^2+1) \quad (c).$$

Over the reals

$$A(x) \equiv (2x-3)(x+\sqrt{2})(x-\sqrt{2})(x^2+1) \quad (b).$$

Over complex numbers we have the complete factorisation

$$(2x-3)(x+\sqrt{2})(x-\sqrt{2})(x+i)(x-i) \quad (a).$$

3. Repeat exercise 2 for $B(x) \equiv x^6+1$.

By the theory of complex numbers the complex roots of $x^6+1 = 0$ are $e^{i\theta}$ where $\theta = \pm\frac{1}{6}\pi, \pm\frac{1}{2}\pi, \pm\frac{5}{6}\pi$. Hence the factorisation is

$$(x+i)(x-i)(x-\tfrac{1}{2}\sqrt{3}-\tfrac{1}{2}i)(x-\tfrac{1}{2}\sqrt{3}+\tfrac{1}{2}i)(x+\tfrac{1}{2}\sqrt{3}+\tfrac{1}{2}i)(x+\tfrac{1}{2}\sqrt{3}-\tfrac{1}{2}i) \quad (a).$$

Over the reals we must group the factors in conjugate pairs, obtaining

$$(x^2+1)(x^2-\sqrt{3}x+1)(x^2+\sqrt{3}x+1) \quad (b).$$

Over the rationals or integers we need to group the last two quadratic factors together, thus

$$(x^2+1)(x^4-x^2+1) \quad (c) \text{ and } (d).$$

4. Find the H.C.F. of the polynomials x^5+1 and $x^5+x^4-2x^3-2x^2+x+1$ and express it in the form $SA+TB$. Factorise each polynomial completely, working throughout over residues modulo 5.

$$x^5+x^4-2x^3-2x^2+x+1 = (x^5+1)+x^4-2x^3-2x^2+x,$$
$$x^5+1 = (x^4-2x^3-2x^2+x)(x+2)+x^3+3x^2-2x+1,$$
$$x^4-2x^3-2x^2+x = (x^3+3x^2-2x+1)x,$$

since $-2 = +3$ (modulo 5). Hence the H.C.F. is x^3+3x^2-2x+1, which is x^3+3x^2+3x+1 or $(x+1)^3$.

$$(x+1)^3 = (x^5+1)-(x+2)(x^4-2x^3-2x^2+x)$$
$$= (x+3)(x^5+1)-(x+2)(x^5+x^4-2x^3-2x^2+x+1).$$

Since $(x+1)^3$ is a factor of both polynomials we may divide out and obtain

$$x^5+1 = (x+1)^3(x^2+2x+1) = (x+1)^5,$$
$$x^5+x^4-2x^3-2x^2+x+1 = (x+1)^3(x^2-2x+1)$$
$$= (x+1)^3(x-1)^2.$$

Exercises 7A

1. What are the units of the polynomials over the residues modulo p, where p is a prime?

Find Q and R such that $A = BQ+R$ and $d(R)<d(B)$ for the pairs of polynomials in **2–6** (all over the rationals):

2. $A \equiv 2x^3+5x^2+2,\qquad B \equiv x+1.$

3. $A \equiv x^7,\qquad B \equiv x^2-x+1.$

4. $A \equiv x^5+x^4+3x^2+2x+3,\qquad B \equiv 3x^2.$

5. $A \equiv \frac{1}{2}x^6+\frac{1}{3}x^2-2,\qquad B \equiv \frac{1}{4}x^2+x.$

6. $A \equiv x^{12}-1,\qquad B \equiv x^2-1.$

Use Euclid's algorithm to find the H.C.F. of the pairs of polynomials **7–11** and express it in the form $SA+TB$:

7. $x^5-x;\quad x^4+x^3-x-1.$

8. $x^5+3x^4+2x^3+4x^2+4x+1$; $x^3+8x^2+16x+5$.

9. x^7-x; x^4+3x^3+15. **10.** $x^{14}-2x^9+x^5-2$; $x^{12}-2x^7+x^5-2$.

11. $2x^5+5x^3+2x$; $2x^3-x^2+x-\frac{1}{2}$.

Factorise the polynomials **12–18** into irreducible factors over
(*a*) the complex numbers; (*c*) the rationals;
(*b*) the real numbers; (*d*) the integers:

12. $2x^3-3x^2+x$. **13.** x^3-2. **14.** x^4-1.

15. $x^3+8x^2+16x+5$. **16.** $2x^3-3x^2+2x-3$.

17. $x^4-x^3+x^2-2x-2$. **18.** x^4+1.

19. Find all the rational roots of the equation

$$x^4+\tfrac{1}{6}x^3+\tfrac{14}{3}x^2+\tfrac{5}{6}x-\tfrac{5}{3} = 0.$$

20. Find *all* the roots of the equation

$$x^4-\tfrac{1}{15}x^3-\tfrac{32}{15}x^2+\tfrac{2}{15}x+\tfrac{4}{15}=0.$$

21. Show that the set of all polynomials of the form $SA+TB$, where A and B are given polynomials, consists of *all* multiples of the H.C.F. of A and B (repeated of course).

22. Prove that a quadratic or cubic polynomial which has no zeros is irreducible.

23. Which of the following are irreducible over the rationals: (i) x^3+7; (ii) $2x^3+5x+1$; (iii) $4x^3+5x-3$; (iv) $4x^3+6x^2-6$?

24. Prove that if U and U' are both units then so is UU'.

25. Give an example to show that if U and U' are units then $U+U'$ is not necessarily one.

Give an example where $U+U'$ *is* a unit.

26. Show that if A and A', B and B' are pairs of associates, then AB and $A'B'$ are associates. Need $A+B$, $A'+B'$ be associates?

27. List all the associates of x^2+3x+2 in the set of polynomials over residues modulo 5.

28. If the polynomial P (over a field) has the property that every other polynomial is either divisible by P or co-prime with it prove that P is irreducible.

29. If $(A, P) = (B, P) = 1$ prove that $(AB, P) = 1$.

Exercises 7B

Prove **1–3** from the abstract definition of a polynomial as (a_i) where all a_i after some a_n are zero, together with the corresponding definitions for sum and product:

1. $(P+Q)+R = P+(Q+R)$.

2. $(PQ)R = P(QR)$.

3. $P(Q+R) = PQ+PR$.

4. Prove theorem 7.5.2.

5. Find the H.C.F. and L.C.M. of the 3 polynomials

$$x^4-4x^3+5x^2-8x+6;$$
$$x^4+4x^3-3x^2+8x-10;$$
$$x^3-3x^2+2x-6.$$

In **6–8** find the H.C.F. and express it in the form $SA+TB$ and also factorise each polynomial completely, working over residues modulo 3:

6. $2x^3+2x^2+x-2;\quad 2x^4+1$.

7. $x^3-1;\quad x^4+2x^3+2x+1$.

8. $x^3+1;\quad x^3+2x$.

9. Repeat the instructions of **6–8** for the following pair of polynomials modulo 5

$$x^3+x^2+x+1;\quad x^4-1.$$

10. Show that the division algorithm (theorem 7.4.2) is true if the coefficients are in any set that possesses commutative addition, subtraction and multiplication and a unity, *provided* B has leading coefficient unity. Deduce the remainder theorem and factor theorem for such a coefficient set.

11. Prove by the method of theorem 7.9.1 that if $F(x, y)$ is a polynomial in the 2 variables x and y then $F(x, y)$ has a factor $(x-y)$ if and only if $F(y, y) \equiv 0$.

12. Let S be a subset of the polynomials over a field such that

(i) $A \in S, B \in S \Rightarrow A \pm B \in S$

(ii) $A \in S \Rightarrow CA \in S$ for any polynomial C.

Prove that either S consists of the zero polynomial alone or it is precisely the set of multiples of a polynomial D, which is a non-zero polynomial of S of lowest degree.

(*Hint.* Use the division algorithm.)

13. What is the remainder when the polynomial $f(x)$ is divided by $(x-a)(x-b)$ (where $a \neq b$), in terms of $f(a)$ and $f(b)$?

(Cambridge Open Scholarship.)

8

VECTORS

8.1. Introduction

This chapter gives a very brief account of some of the properties of vectors, its purpose being to deal with an example of an algebraic structure that differs in important respects from most that we have met so far, and to provide a source for future examples. The emphasis is on the algebraic properties of vectors. It will be noticed that there is no mention of matrices: these form a huge subject in themselves, and one which is quite different in its methods to most of the topics of this book, since in matrix theory we are dealing with individual elements of our sets, while the main interest in abstract algebra is usually in the structure of the set as a whole—it is synthetic whereas matrices and linear algebra are analytic. Much of the work of this chapter will be already familiar to the reader, and he may well omit the parts that he knows thoroughly, although he should be careful to note the structural properties of the sets of vectors.

Many physical (and mathematical) quantities are specified completely by a number, together with some system of units. For example, the length of a line is measured by a number of inches, or centimetres, or feet and inches, while the mass of a body is measured as so many pounds or grams. Other examples are area and volume, temperature, work and energy, speed, electric potential.

For other quantities, however, we need to know not only their magnitude, but also the direction in which they act. Thus to measure velocity we need its magnitude (the speed), measured in terms of a unit such as ft. per sec. or cm. per sec., and also its direction: a velocity of 50 ft. per sec. due north is not the same as one of 50 ft. per sec. due east. Other examples are displacement or movement from one point to another, acceleration, force and momentum, electric and magnetic intensity.

Quantities that are specified completely by a number are

called *scalars*, while those that possess magnitude and direction are known as *vectors*, and are usually written in bold type—thus: **v**.

Since magnitude and direction are sufficient to specify a vector it may be represented by a line drawn in space. For the moment we will consider all vectors as being in two dimensions in the plane of the paper: we will later generalise to three (and more) dimensions. Then a (two-dimensional) vector may be represented by a line drawn in the paper, thus

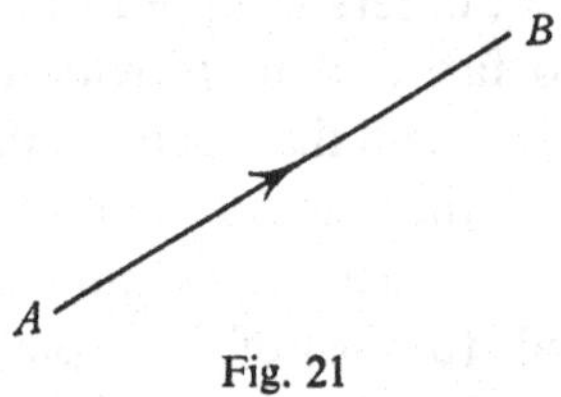

Fig. 21

We say that the vector is represented by the displacement from A to B, and denote it by **AB**. Note that we use bold type for **AB**, otherwise we would mean the length (the scalar) AB. Note also that **BA** is not the same as **AB**, since it is in the opposite direction. We always write the starting point first—**AB** means a displacement or movement *from A to B*. We often put an arrow on the line also, as in figure 21.

The length or magnitude of a vector **a** is often called the *modulus* of **a** and written a, or $|\mathbf{a}|$. (This of course reminds us of complex numbers, which are indeed a type of vector as will be seen later.)

While some physical quantities are specified completely by magnitude and direction others depend also on their line of action. The most important case is that of force, whose effect depends on the line along which it acts, although it may be considered to act at any point of the line. Such vectors are said to be localised in a line, or to be *line vectors*. A vector which is not localised is called a *free vector*. It is found convenient mathematically to treat all vectors as free: in mechanics or in any application where the vectors are in fact localised we must take the localisation into account: in the case of forces this leads to the ideas of moment and couple.

The above explains what we usually mean by vectors in intuitive terms, but hardly gives a satisfactory *definition*. The best elementary algebraic definition is in terms of co-ordinates (a more abstract definition is given in volume 2, when we deal with vector spaces). If we treat all vectors as starting at the origin then we may represent them fully by the final point (e.g. in figure 21 if A is the origin we may think of the vector as represented by the point B). Thus the vector is defined by the co-ordinates of this final point.

For the moment we restrict ourselves to two-dimensional vectors, but the work is easily extended later to more dimensions. In the next section we will give our co-ordinate definition and discuss the algebra of vectors, while in §8.3 we give geometrical illustrations of the basic laws.

8.2. Definition and algebra of vectors in two dimensions

We define a vector (in two dimensions) to be a pair of numbers (x, y). The numbers, or *scalars*, x and y may be real or complex or indeed elements of any field, and the set of vectors over this field consists of all possible ordered pairs (x, y). Two vectors are equal if and only if both co-ordinates x and y are equal.

Addition of vectors

If $\mathbf{a} = (a_1, a_2)$ and $\mathbf{b} = (b_1, b_2)$ we define $\mathbf{a}+\mathbf{b}$ to be the vector (a_1+b_1, a_2+b_2), and note that this is in the set.

Theorem 8.2.1. *Addition of vectors is commutative, i.e.*

$$\mathbf{a}+\mathbf{b} = \mathbf{b}+\mathbf{a}.$$

This follows at once since addition of the scalar co-ordinates is commutative.

Theorem 8.2.2. *Addition of vectors is associative, i.e.*

$$(\mathbf{a}+\mathbf{b})+\mathbf{c} = \mathbf{a}+(\mathbf{b}+\mathbf{c}).$$

This again follows from the associativity of the co-ordinates.

Subtraction of vectors

The zero vector $\mathbf{0}$ is the vector $(0, 0)$, and we note that

$$\mathbf{0}+\mathbf{a} = \mathbf{a}+\mathbf{0} = \mathbf{a}.$$

The negative of $\mathbf{a} = (a_1, a_2)$ is $(-a_1, -a_2)$ and is written $-\mathbf{a}$. Then $\mathbf{a}+-\mathbf{a} = -\mathbf{a}+\mathbf{a} = \mathbf{0}$.

To subtract $\mathbf{b}$ from $\mathbf{a}$ we either add $-\mathbf{b}$ to $\mathbf{a}$ or, what amounts to the same thing, form the vector (a_1-b_1, a_2-b_2). Of course $\mathbf{b}-\mathbf{a} = -(\mathbf{a}-\mathbf{b})$.

Multiplication of a vector by a scalar

We write $\mathbf{a}+\mathbf{a} = 2\mathbf{a}$, $2\mathbf{a}+\mathbf{a} = 3\mathbf{a}$, etc., and note that $n\mathbf{a} = (na_1, na_2)$ where n is any integer. This leads us to define $\lambda\mathbf{a}$ where λ is *any* scalar (i.e. a real number if our co-ordinates are real numbers, complex if they may be complex, etc.) to be $(\lambda a_1, \lambda a_2)$, and note that this is in our set of vectors.

In order that we may manipulate vectors and scalars with confidence we need to be sure that their product obeys certain basic laws. These are proved in the following theorem. The choice of laws may seem strange (especially (iv)), but it can be shown (see volume 2, *Vector Spaces*) that these are sufficient to ensure that our algebra of vectors behaves as we would expect it to. (They are also all necessary for this.)

Theorem 8.2.3. (i) $\lambda(\mathbf{a}+\mathbf{b}) = \lambda\mathbf{a}+\lambda\mathbf{b}$;
(ii) $(\lambda+\mu)\mathbf{a} = \lambda\mathbf{a}+\mu\mathbf{a}$;
(iii) $\lambda(\mu\mathbf{a}) = (\lambda\mu)\mathbf{a}$;
(iv) $1.\mathbf{a} = \mathbf{a}$.

These are all obvious after a moment's thought. As an example we will prove (i).

$$\begin{aligned}\lambda(\mathbf{a}+\mathbf{b}) &= \lambda(a_1+b_1, a_2+b_2) = (\lambda(a_1+b_1), \lambda(a_2+b_2))\\ &= (\lambda a_1+\lambda b_1, \lambda a_2+\lambda b_2) = (\lambda a_1, \lambda a_2)+(\lambda b_1, \lambda b_2)\\ &= \lambda(a_1, a_2)+\lambda(b_1, b_2) = \lambda\mathbf{a}+\lambda\mathbf{b}.\end{aligned}$$

Note that if we denote (1, 0) by $\mathbf{i}$ and (0, 1) by $\mathbf{j}$ then

$$(x, y) = x\mathbf{i}+y\mathbf{j},$$

and this expression for a vector in terms of $\mathbf{i}$ and $\mathbf{j}$ is unique. Similarly, if we take any two non-zero vectors where one is not a scalar times the other, then every vector can be expressed uniquely in terms of them. Thus if $\mathbf{a} = (1, 2)$ and $\mathbf{b} = (4, 3)$ then $(10, 5) = -2\mathbf{a} + 3\mathbf{b}$, and this expression is unique as may be seen by letting $(10, 5) = \lambda\mathbf{a}+\mu\mathbf{b}$ and solving the resulting simultaneous equations for λ and μ. This corresponds to taking the co-ordinate axes to lie along **a** and **b**, and at the same time changing the scale so that **a** and **b** are unit lengths. We are in effect finding the components of (10, 5) in the two directions **a** and **b**, in terms of multiples of **a** and **b**. (Note that **a** and **b** are not at right angles.)

8.3. The algebra of vectors illustrated geometrically

A two-dimensional vector may be represented by a line in a plane, such that all equal and parallel lines represent the same vector. Thus if $\mathbf{a} = (1, 2)$ then if A is the point (1, 2), **a** may be represented by any line equal and parallel to **OA**.

It is easy to give geometrical interpretations of the operations of addition and multiplication by a scalar, and it is interesting to illustrate the basic laws by this means.

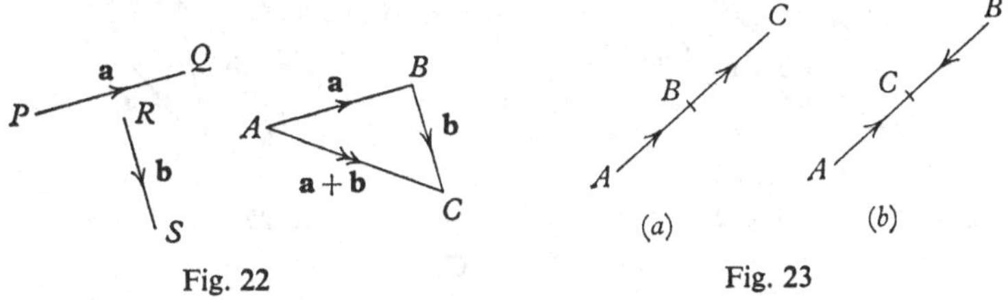

Fig. 22

Fig. 23

Addition

If two vectors **a** and **b** are represented by **PQ** and **RS**, to obtain $\mathbf{a}+\mathbf{b}$ we move them parallel to themselves so that the initial point of **b** corresponds to the final point of **a**; thus **a** is represented by **AB** in figure 22 and **b** by **BC**. Then $\mathbf{a}+\mathbf{b}$ is represented by **AC**.

By the properties of congruent triangles the sum is independent of the lines of action of the summands—its magnitude and direction depend only on their magnitudes and directions.

If the vectors are parallel the triangle becomes degenerate as in figures 23(*a*) and (*b*).

That this definition of addition follows from our co-ordinate definition is almost immediate, by taking A as the origin O, and noting that the projection of **AC** on either axis is equal to the sum of the projections of **AB** and **BC**, so that the co-ordinates of the vector **AC** *are* equal to the sums of those of **AB** and **BC**. It is also the obvious meaning to be given to the sum of two displacements, and it can be shown experimentally that this triangle law of addition is true for the resultant of two forces or velocities, so that the resultant is represented merely as the sum of the component vectors.

The Commutative Law is illustrated in figure 24, where $ABCD$ is a parallelogram, and

$$\mathbf{AC} = \mathbf{AB}+\mathbf{BC} = \mathbf{AD}+\mathbf{DC} = \mathbf{a}+\mathbf{b} = \mathbf{b}+\mathbf{a}.$$

If the vectors **a** and **b** are parallel the result follows from the commutativity of real numbers.

The Associative Law is illustrated in figure 25.

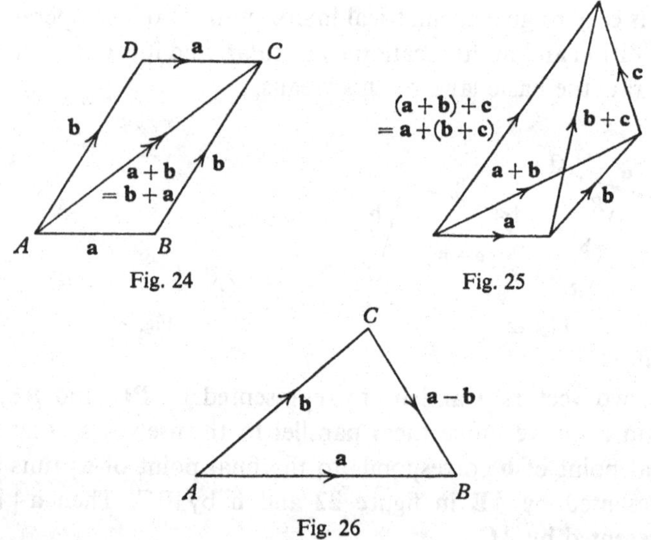

Fig. 24

Fig. 25

Fig. 26

Subtraction

If **a** and **b** are represented by **AB** and **AC**, respectively, then **CB** represents $\mathbf{a}-\mathbf{b}$. For $-\mathbf{b}$ is clearly represented by **CA**, and so $\mathbf{a}-\mathbf{b} = -\mathbf{b}+\mathbf{a} = \mathbf{CA}+\mathbf{AB} = \mathbf{CB}$.

Multiplication by a scalar

If **a** is represented by **AB** and $\lambda > 0$, $\lambda\mathbf{a}$ is represented by a line parallel to **AB** and equal in magnitude to λAB. This of course gives $-\lambda\mathbf{a}$ as represented by a line parallel to **AB** but in the opposite direction, and λAB in magnitude. The definition is clearly consistent with the idea of $2\mathbf{a} = \mathbf{a}+\mathbf{a}$, etc. If λ is zero then $\lambda\mathbf{a}$ is the zero vector (represented by the line segment with both ends coincident, i.e. by the point itself).

Parts (ii), (iii) and (iv) of theorem 8.2.3 are obvious, and part (i) is illustrated in figures 27 and 28. In both these figures **AB** and **AD** represent $\mathbf{a}$ and $\lambda\mathbf{a}$ respectively, where λ is positive in figure 27 and negative in figure 28. **BC** and **DE** represent $\mathbf{b}$ and $\lambda\mathbf{b}$. Then $\mathbf{a}+\mathbf{b}$, $\lambda\mathbf{a}+\lambda\mathbf{b}$ are represented by **AC** and **AE**. But triangles ABC and ADE are similar and so ACE are collinear and $AE = \lambda AC$. Hence $\mathbf{AE} = \lambda\mathbf{AC}$, i.e. $\lambda(\mathbf{a}+\mathbf{b}) = \lambda\mathbf{a}+\lambda\mathbf{b}$.

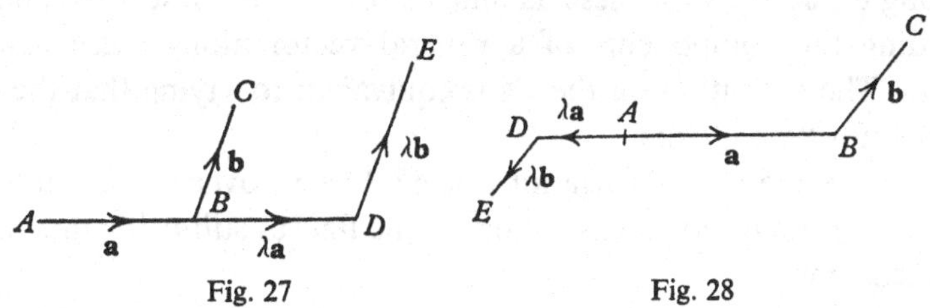

Fig. 27 Fig. 28

In the above geometrical representation of vectors we have assumed that the co-ordinates (the scalars) are real numbers. In other cases we would need to be careful about our geometrical figures—for two-dimensional vectors over the complex numbers we need four dimensions in which to represent them (since the complex numbers form a two-dimensional space).

8.4. Vectors in three dimensions

We have so far dealt entirely with two-dimensional vectors, but vectors in three dimensions may be defined in the same way. We define a three-dimensional vector as an ordered triad of numbers (x, y, z), where the co-ordinates may be real, complex or in any field. We define $\mathbf{a}+\mathbf{b}$, where $\mathbf{a} = (a_1, a_2, a_3)$ and $\mathbf{b} = (b_1, b_2, b_3)$ to be $(a_1+b_1, a_2+b_2, a_3+b_3)$, and the Commutative and Associative Laws follow as for the two-dimensional case. $-\mathbf{a}$ is $(-a_1, -a_2, -a_3)$ and $\mathbf{0}$ is $(0, 0, 0)$. $\lambda\mathbf{a}$ is defined to be $(\lambda a_1, \lambda a_2, \lambda a_3)$ and the laws of theorem 8.2.3 follow.

If we denote the vectors $(1, 0, 0)$, $(0, 1, 0)$ and $(0, 0, 1)$ by $\mathbf{i}$, $\mathbf{j}$, and $\mathbf{k}$, then any vector $\mathbf{a}$ may be represented uniquely in the form $a_1\mathbf{i}+a_2\mathbf{j}+a_3\mathbf{k}$. Similarly, if we take any three vectors such that no one is a linear combination of the other two (i.e. if we take $\mathbf{e}_1$, $\mathbf{e}_2$, $\mathbf{e}_3$ it must not be possible to choose numbers λ and μ such that $\mathbf{e}_1 = \lambda\mathbf{e}_2+\mu\mathbf{e}_3$, and similarly for the other **e**'s: put another way, if $\lambda_1\mathbf{e}_1+\lambda_2\mathbf{e}_2+\lambda_3\mathbf{e}_3 = \mathbf{0}$ then $\lambda_1, \lambda_2, \lambda_3$ are all zero) then any vector is expressible uniquely in terms of them. For example, if $\mathbf{e}_1 = (2, 1, 0)$, $\mathbf{e}_2 = (1, -3, 2)$ and $\mathbf{e}_3 = (0, 0, 1)$, then $(1, 11, -5) = 2\mathbf{e}_1-3\mathbf{e}_2+\mathbf{e}_3$, and this is unique, as may be seen by letting the coefficients be α, β, γ and solving the three resulting simultaneous equations for α, β, and γ. As in the two-dimensional case we are in effect taking new co-ordinate axes along $\mathbf{e}_1$, $\mathbf{e}_2$, $\mathbf{e}_3$, with these as unit vectors along these axes, and finding the components of a general vector along these new axes. The condition on the **e**'s is equivalent to saying that they must not be coplanar.

The geometrical illustrations of §8.3 carry over at once into the three-dimensional case, most of the figures still being in two dimensions.

8.5. The use of vectors in Cartesian geometry

Some problems in elementary Euclidean geometry lend themselves to a vector solution; when this exists it is usually the best method of solving the problem, replacing as it does a set of two or more co-ordinates by a single vector. In this section we will give some of the theory, particularly that involving centroids, but to gain a good idea of the methods used the reader should study the worked exercises at the end of the chapter.

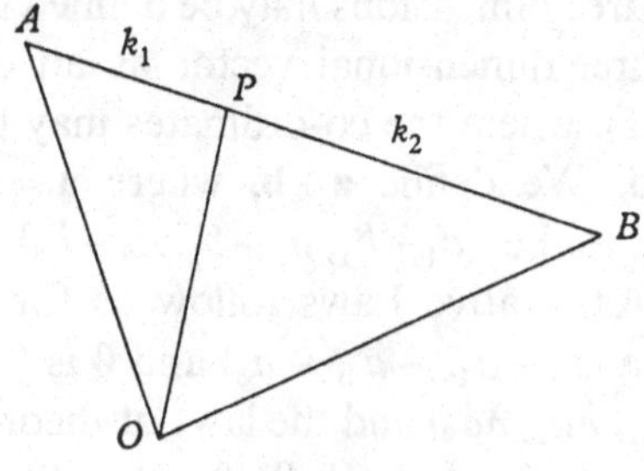

Fig. 29

Theorem 8.5.1 (*see figure* 29). *If P is the point dividing the line AB in the ratio $k_1:k_2$ and O is any point, then*

$$\mathbf{OP} = \frac{k_2\mathbf{OA}+k_1\mathbf{OB}}{k_1+k_2}.$$

$$k_2\mathbf{AP} = k_1\mathbf{PB},$$

i.e.

$$k_2(\mathbf{OP}-\mathbf{OA}) = k_1(\mathbf{OB}-\mathbf{OP}),$$

$$(k_1+k_2)\mathbf{OP} = k_2\mathbf{OA}+k_1\mathbf{OB}.$$

Centroids

The centroid of masses m_i at A_i, where $i = 1, 2, \ldots, n$ and $\Sigma m_i \neq 0$, is defined to be the point G such that, if O is any point,

$$\mathbf{OG} = \frac{\Sigma m_i \mathbf{OA}_i}{\Sigma m_i}.$$

It is unique by the next theorem.

The centroid is the centre of gravity of the masses. The centroid of unit masses at A_i is often known as the centroid of the points.

Theorem 8.5.2. *The centroid is independent of O.*

Suppose the centroids obtained from two points O and O' are G and G'. Then

$$(\Sigma m_i)\mathbf{OG} = \Sigma m_i \mathbf{OA}_i \quad \text{and} \quad (\Sigma m_i)\mathbf{O'G'} = \Sigma m_i \mathbf{O'A}_i.$$

Hence

$$(\Sigma m_i)(\mathbf{OG}-\mathbf{O'G'}) = \Sigma m_i(\mathbf{OA}_i-\mathbf{O'A}_i) = \Sigma m_i \mathbf{OO'}.$$

But $\quad \Sigma m_i \neq 0, \quad$ and so $\quad \mathbf{OG}-\mathbf{O'G'} = \mathbf{OO'}.$

Thus $\quad \mathbf{OG} = \mathbf{OO'}+\mathbf{O'G'} = \mathbf{OG'} \quad$ and $\quad G = G'.$

Note. It is easily seen from the definition that to obtain the centroid of the masses m_i at A_i and n_j at B_j we may first find the centroids of the masses m_i and n_j separately and then consider the masses Σm_i at G_1 and Σn_j at G_2, where G_1 and G_2 are these centroids. Note also by theorem 8.5.1, that the centroid of m_1 at A_1 and m_2 at A_2 is the point dividing A_1A_2 in the ratio $m_2:m_1$.

Theorem 8.5.3. *Properties of a triangle by centroids.*

(i) *The medians of a triangle ABC are concurrent in G, the centroid of unit masses at A, B and C, and G trisects each median.*

(ii) *The altitudes of a triangle ABC are concurrent in H, the centroid of masses tan A, tan B, tan C at A, B and C respectively.*

(iii) *The perpendicular bisectors of the sides of a triangle are concurrent in O, the centroid of masses* sin $2A$, sin $2B$ *and* sin $2C$ *at the vertices.*

(iv) *The internal bisectors of the angles are concurrent in I, the centroid of masses* sin A, sin B, sin C *at the vertices.*

(i) The centroid of unit masses at B and C is A', the mid-point of BC. Hence the centroid of unit masses at A, B and C is the point G which trisects the median $A'A$ (using the note above). Similarly, the centroid trisects the other medians, and hence the three medians are concurrent.

(ii) Let AD be an altitude. Then

$$BD{:}DC = AD \cot B{:}AD \cot C = \tan C{:}\tan B.$$

Thus the centroid of tan B at B and tan C at C is at D, and so the centroid of the three given masses lies on AD, and similarly on the other two altitudes.

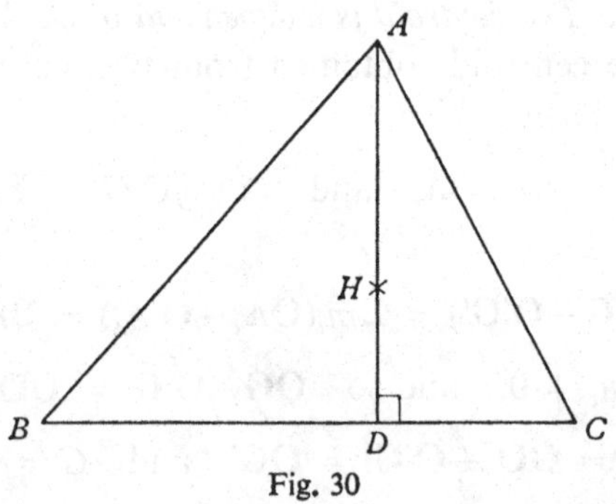

Fig. 30

(iii) Let the perpendicular bisectors of AB and AC meet in O, and produce AO to meet BC at U. Then

$$\begin{aligned} BU{:}UC &= \triangle ABU{:}\ \triangle ACU \\ &= \tfrac{1}{2}AU.AB.\sin O\hat{A}B{:}\tfrac{1}{2}AU.AC.\sin O\hat{A}C \\ &= AB.\cos C{:}AC.\cos B \end{aligned}$$

(for $OA = OB = OC$ and so $A\hat{O}B = 2C$, being the angle at the centre of the circle ABC. Hence $A\hat{O}N = C$ and similarly $A\hat{O}M = B$)

$$= \sin C.\cos C : \sin B.\cos B \quad \text{by the sine rule}$$

$$= \sin 2C : \sin 2B.$$

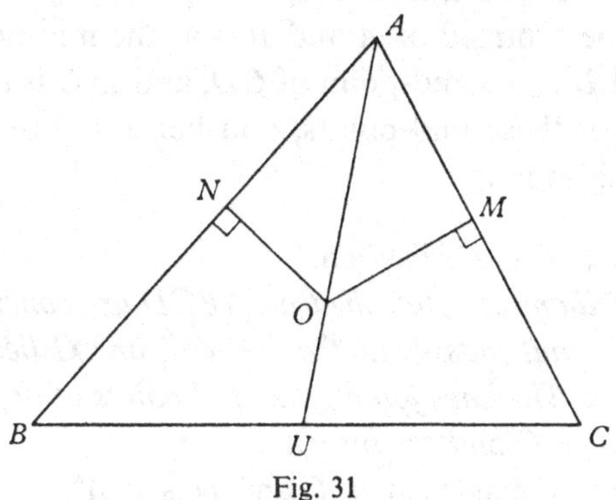

Fig. 31

Hence U is the centroid of $\sin 2B$ at B and $\sin 2C$ at C, and so the centroid of the three masses lies on AU, and so on AO. Similarly, it lies on BV and CW, defined similarly to AU, and thus is O, since O is on the perpendicular bisector of BC (as $OA = OB = OC$) and so on BV and CW.

(iv) Let the bisector of A meet BC in U. Then

$$BU : UC = AB : AC = \sin C : \sin B.$$

Hence U is the centroid of $\sin B$ at B and $\sin C$ at C, and so the centroid of the three masses lies on AU, and similarly on the other two bisectors.

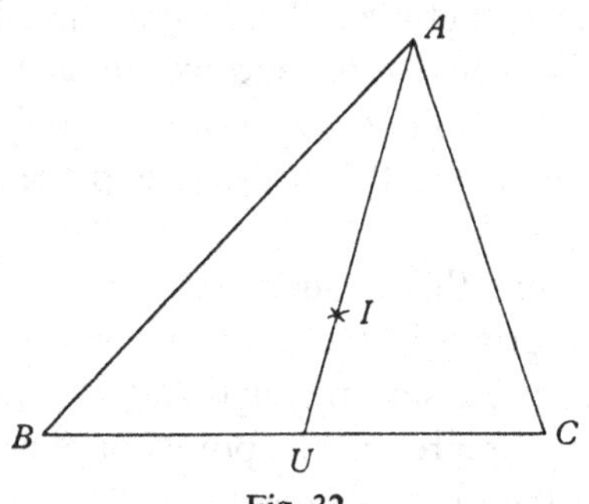

Fig. 32

Theorem 8.5.4. *The quadrilateral.*

The lines joining the mid-points of opposite sides of a quadrilateral ABCD bisect one another in G, the centroid of the points A, B, C, D, and the line joining the mid-points of the diagonals also passes through G and is bisected by it.

The proof is almost immediate, by considering the masses in pairs. Thus the centroid of A and B is at the mid-point of AB, that of C and D is the mid-point of CD, and so G is the centroid of 2 at each of these mid-points, and hence G bisects the line joining the mid-points.

Theorem 8.5.5. *The tetrahedron.*

The four medians of a tetrahedron ABCD are concurrent in G, the centroid of unit masses at the vertices, and G lies $\frac{1}{4}$ the way up each median. The lines joining the mid-points of opposite edges also pass through G and are bisected by it.

The centroid of masses at B, C and D is at A', the centroid of the triangle BCD, and so G is the centroid of 3 at A' and 1 at A. The first part follows.

The second part is proved as in theorem 8.5.4 by taking the masses in pairs.

8.6. Projective geometry

In two-dimensional projective geometry we use three homogeneous co-ordinates to specify a point (x, y, z) and are interested only in the ratios $x:y:z$. Thus if $\mathbf{a} = (x, y, z)$, $\mathbf{a}$ and $\lambda\mathbf{a}$ represent the same point. $\mathbf{0}$ does not represent any point.

Thus we may obtain the points in the projective plane by an equivalence relation between three-dimensional vectors. $\mathbf{a}R\mathbf{b}$ if $\mathbf{b} = \lambda\mathbf{a}$ where λ is non-zero, and we exclude the zero vector completely. This relation is obviously an equivalence relation and gives the points of the projective plane as equivalence classes.

We may represent this geometrically by considering a unit sphere centre the origin in three-dimensional space. Each line in three-dimensions passing through O gives us one projective point. Thus we may represent the points of our projective plane by the points on the upper hemisphere of the unit sphere, with points on its edge (for which $z = 0$) identified so that $(x, y, 0)$ and $(-x, -y, 0)$ give rise to the same point. This is the idea behind the topological 'projective plane'—the edges of the hemisphere are 'sewn' together antipodally as above to give a surface, which unfortunately must cross itself when exhibited in three dimensions only.

Projective space of higher dimension may be defined similarly —for n dimensions we need $n+1$ co-ordinates and form an equivalence relation between them.

8.7. *n*-dimensional vectors

n-dimensional vectors are defined as ordered n-tuples of numbers $(a_1, a_2, \ldots, a_n)$, where the a_i's may be in any field. Addition and multiplication by a scalar are defined exactly as before and the usual rules verified.

The vector given by all co-ordinates zero except the ith, which is 1, is denoted by $\mathbf{e}_i$, and any vector may be expressed in the form $a_1\mathbf{e}_1 + a_2\mathbf{e}_2 + \ldots + a_n\mathbf{e}_n$. The set $\mathbf{e}_1, \mathbf{e}_2, \ldots \mathbf{e}_n$ is called a *base* for the vectors, and no $\mathbf{e}_i$ is expressible in terms of the others.

Four-dimensional vectors are used in the theory of relativity, and in electricity and magnetism.

8.8. Multiplication of vectors

The two basic operations that we have applied to vectors are addition (including subtraction) and multiplication by a scalar. In this section we investigate the possibility of multiplying two vectors.

In general there is no satisfactory definition of product of vectors. Such a definition should obey at least some of the usual laws of algebra, and it turns out that this is not always possible. There are, however, certain types of product that are useful in special cases, and some of these will be mentioned below.

We will indicate first why a general product will not work. By analogy with the sum the natural definition of product is that (using two-dimensional vectors for simplicity), if $\mathbf{a} = (a_1, a_2)$ and $\mathbf{b} = (b_1, b_2)$, then $\mathbf{ab} = (a_1b_1, a_2b_2)$. But if we adopt this definition we have the possibility of the product of two non-zero vectors being zero. For example, if $\mathbf{a} = (0, a)$ and $\mathbf{b} = (b, 0)$, then $\mathbf{ab} = (0, 0) = \mathbf{0}$, even though a and b, and so $\mathbf{a}$ and $\mathbf{b}$, are not zero. This is a serious defect, since not only does it preclude the possibility of division always being possible but it also prevents us doing much simple multiplicative algebra, as the property that $ab = 0 \Rightarrow$ either $a = 0$ or $b = 0$ is fundamental in the usual algebraic work, except in the theory of rings.

We thus need more elaborate definitions of product, and it is not often possible to find ones that satisfy enough of the fundamental laws to be useful.

Complex numbers

Referring to §5.3 we see that the sum of complex numbers is defined in exactly the same way as the sum of vectors: that $(a_1, a_2)+(b_1, b_2) = (a_1+b_1, a_2+b_2)$; and that multiplication by a scalar is the same: $\lambda(a, b) = (\lambda, 0)(a, b) = (\lambda a, \lambda b)$. Hence, complex numbers are a special case of two-dimensional vectors, being characterised by the fact that their product is also defined. This product (*not* a natural one from the vector point of view) satisfies all the fundamental algebraic laws. The geometrical problems that may be solved by vectors can be thought of alternatively as problems about the Argand diagram, and the multiplicative structure of complex numbers can sometimes help us even more than the additive structure (which is common to all vectors). For an example of this see Worked exercise no. 1.

Quaternions

It is not possible to put a satisfactory multiplicative structure on three-dimensional vectors (satisfactory from the algebraic point of view—structures exist that are very useful in other ways, as we see below), but we can do so on vectors in four dimensions, by treating them as quaternions as in §5.5. Apart from the Commutative Law these obey all the fundamental laws.

Extending this, multiplication may be defined similarly on vectors in 8 dimensions, giving the so-called Cayley numbers, and in fact in 2^n dimensions where n is any positive integer. These, however, lose more of the fundamental laws.

Inner or scalar product

Our definition of vectors in terms of co-ordinates gives no meaning to the length of a vector. We have indeed used the term, but have thought of it intuitively in terms of geometry. It is, however, possible to define length and angle abstractly, and we give an outline of the method.

We define the *inner* or *scalar product* of two n-dimensional vectors $\mathbf{a} = (a_1, a_2, \ldots, a_n)$ and $\mathbf{b} = (b_1, b_2, \ldots, b_n)$ to be the scalar $a_1 b_1 + a_2 b_2 + \ldots + a_n b_n$ and denote it by $\mathbf{a}.\mathbf{b}$. Dealing with the case of vectors over the real numbers only, we define the length of $\mathbf{a}$, denoted by $|\mathbf{a}|$, to be $\sqrt{(\mathbf{a}.\mathbf{a})}$ or

$$\sqrt{(a_1^2 + a_2^2 + \ldots + a_n^2)},$$

where we take the positive square root and note that it always exists since $a_1^2 + a_2^2 + \ldots + a_n^2$ is a positive real number. This definition of course corresponds to the geometrical idea of length (at any rate in two or three dimensions and, by analogy, in higher dimensions also) by Pythagoras's theorem.

To define angle, we note that in Cartesian geometry in a plane, if $\mathbf{a} = (a_1, a_2)$ and $\mathbf{b} = (b_1, b_2)$, then

$$a_1 b_1 + a_2 b_2 = |\mathbf{a}|\,|\mathbf{b}| \cos \theta$$

where θ is the angle between $\mathbf{a}$ and $\mathbf{b}$. [This may easily be proved for, if $\mathbf{a}$ and $\mathbf{b}$ make angles α and β with the x-axis,

$$a_1 = |\mathbf{a}| \cos \alpha \quad a_2 = |\mathbf{a}| \sin \alpha \quad b_1 = |\mathbf{b}| \cos \beta \quad b_2 = |\mathbf{b}| \sin \beta$$

and so the

$$\begin{aligned} \text{L.H.S.} &= |\mathbf{a}|\,|\mathbf{b}|(\cos \alpha \cos \beta + \sin \alpha \sin \beta) \\ &= |\mathbf{a}|\,|\mathbf{b}|(\cos(\alpha - \beta)).] \end{aligned}$$

A similar result holds in three dimensions, and suggests the general definition of the angle between $\mathbf{a}$ and $\mathbf{b}$ as being given by

$$\cos \theta = \frac{\mathbf{a}.\mathbf{b}}{(\mathbf{a}.\mathbf{a})^{\frac{1}{2}}(\mathbf{b}.\mathbf{b})^{\frac{1}{2}}}.$$

(This may be proved to lie between -1 and $+1$ and so always gives a real angle, unique if we restrict θ to the range $0 \leqslant \theta \leqslant \pi$.)

Thus scalar product is basic to the idea of length and angle. It is not of great importance for our present purposes however, since it is not itself a vector. Its main *practical* use is in connection with resolved parts: the resolved part of **a** in the direction of the unit vector **e** is **a**.**e**.

Vector product

This is denoted by $\mathbf{a} \times \mathbf{b}$ or $\mathbf{a} \wedge \mathbf{b}$ and is defined in three dimensions to be the vector $(a_2 b_3 - a_3 b_2, a_3 b_1 - a_1 b_3, a_1 b_2 - a_2 b_1)$. It is a vector but satisfies few of the basic laws. Thus

$$\mathbf{a} \wedge \mathbf{b} = -\mathbf{b} \wedge \mathbf{a},$$

while the Associative Law is not true in general (see theorem 10.7.1). The Distributive Law does, however, hold:

$$\mathbf{a} \wedge (\mathbf{b}+\mathbf{c}) = \mathbf{a} \wedge \mathbf{b} + \mathbf{a} \wedge \mathbf{c}.$$

The vector product is a vector at right angles to both **a** and **b** having magnitude $|\mathbf{a}|\ |\mathbf{b}| . \sin\theta$ where **a** and **b** are at an angle θ. (It is fundamental in the theory of moments: the vector moment of a vector **a** whose position vector is **r** is $\mathbf{r} \wedge \mathbf{a}$, and the scalar moment about a line through the origin parallel to the unit vector **b** is $(\mathbf{r} \wedge \mathbf{a}).\mathbf{b}$.)

The analogue of the vector product in four dimensions has *six* components and so may no longer be thought of as a vector, at least in four dimensions. In any number of dimensions it may be put in tensor or matrix form.

Outer product

The set of n^2 quantities $a_i b_j$ for $i = 1, 2, 3, \ldots, n$ and $j = 1, 2, 3, \ldots, n$ form the so-called *outer product* of **a** and **b**. It may be defined in any number of dimensions, being in fact a tensor or matrix. Its use is more restricted than the two previous products. It is sometimes known as the *open product*.

8.9. More examples of vectors

Many objects may be thought of as vectors, and the laws of vectors applied. We have already met the basic physical vectors, and complex numbers, and some other examples are given below.

Polynomials

The polynomials of degree $\leqslant n$ form a set of vectors in $n+1$ dimensions, with the usual laws of vector addition and scalar multiplication. There is no product, since the usual polynomial product increases the degree of the polynomials and so the dimension of the vectors. *All* polynomials may, however, be thought of as vectors in an infinite number of dimensions (special in the sense that all coefficients or co-ordinates after a finite number are zero) and then product is perfectly well defined and obeys the basic laws, excepting those of division.

Sets of marks

If a group of pupils each have marks awarded for a number of pieces of work these may be thought of as the co-ordinates of n-dimensional vectors, where there are n sets of marks obtained.

Intelligence testing

This is similar to the above. If intelligence ratings are given in several subjects or types of tests then the results may be tabulated in vector form, and information may be obtained by thinking of them in geometrical ways. For example, if the marks in two allied tests are x and y then the lines $x+y =$ constant will give all the people who are of the same standard in both these tests combined.

Worked exercises

1. X, Y, Z are the centres of squares described externally on the sides BC, CA, AB of a triangle. Prove that AX, YZ are perpendicular and of equal length. (Cambridge Open Scholarship.)

We will use complex numbers (as vectors with a multiplicative structure), treating the plane as the Argand diagram with some unspecified origin O and denoting points such as A by the complex numbers a, etc.

We wish to prove that the vectors **AX** and **YZ** have equal moduli and amplitudes differing by a right angle, i.e. that $x-a = i(z-y)$. We start by finding x, y, z in terms of a, b, c. (Note that by taking an arbitrary origin we keep symmetry and thus need only find x, writing down symmetrical expressions for y and z.)

BX has modulus equal to $1/\sqrt{2}$ times that of **BC**, while its amplitude equals $am(\mathbf{BC})+\frac{1}{4}\pi$.

Hence $$x-b = (1/\sqrt{2})\, e^{\frac{1}{4}\pi i}(c-b)$$
$$= 1/\sqrt{2}[(1/\sqrt{2})+(1/\sqrt{2})i](c-b)$$
$$= \tfrac{1}{2}(1+i)(c-b),$$
therefore $$x = \tfrac{1}{2}(1+i)c+\tfrac{1}{2}(1-i)b.$$
Similarly $$y = \tfrac{1}{2}(1+i)a+\tfrac{1}{2}(1-i)c,$$
$$z = \tfrac{1}{2}(1+i)b+\tfrac{1}{2}(1-i)a.$$
Hence $$z-y = -ia+\tfrac{1}{2}(1+i)b-\tfrac{1}{2}(1-i)c$$
$$= i[-a+\tfrac{1}{2}(1-i)b+\tfrac{1}{2}(1+i)c]$$
$$= i(x-a).$$

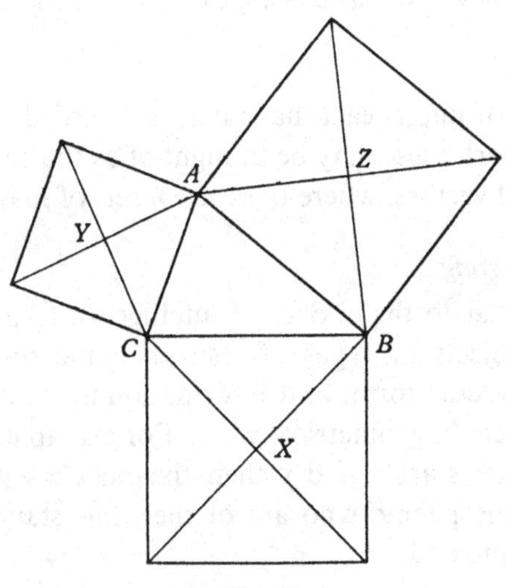

Fig. 33

2. Express the vector (2, 5) in terms of $\mathbf{a} = (-\frac{3}{5}, \frac{4}{5})$ and $\mathbf{b} = (\frac{4}{5}, \frac{3}{5})$ and verify that if it is equal to $x'\mathbf{a}+y'\mathbf{b}$, then $x'^2+y'^2 = 2^2+5^2$.

If $(2, 5) = x'\mathbf{a}+y'\mathbf{b}$ we have
$$2 = -\tfrac{3}{5}x'+\tfrac{4}{5}y',$$
$$5 = \tfrac{4}{5}x'+\tfrac{3}{5}y'.$$
Solving in the usual way
$$x' = \tfrac{14}{5}, \quad y' = \tfrac{23}{5}.$$
$$x'^2+y'^2 = \tfrac{196}{25}+\tfrac{529}{25} = \tfrac{725}{25} = 29 = 2^2+5^2.$$
(**a** and **b** are unit vectors which are at right angles. We are in effect rotating our axes so that **a** and **b** are the new unit vectors along the axes, and the modulus of (2, 5) is of course unchanged, and is $\sqrt{(x'^2+y'^2)}$ in terms of its new co-ordinates.)

3. If G is the mid-point of AB and G' is the mid-point of $A'B'$ prove that $\mathbf{AA'}+\mathbf{BB'} = 2\mathbf{GG'}$.

$$\mathbf{AA'} = \mathbf{AG}+\mathbf{GG'}+\mathbf{G'A'},$$
$$\mathbf{BB'} = \mathbf{BG}+\mathbf{GG'}+\mathbf{G'B'}.$$

Hence $$\mathbf{AA'}+\mathbf{BB'} = \mathbf{AG}+\mathbf{BG}+2\mathbf{GG'}+\mathbf{G'A'}+\mathbf{G'B'}.$$

But $$\mathbf{AG} = -\mathbf{BG} \quad \text{and} \quad \mathbf{G'A'} = -\mathbf{G'B'},$$

and so $$\mathbf{AA'}+\mathbf{BB'} = 2\mathbf{GG'}.$$

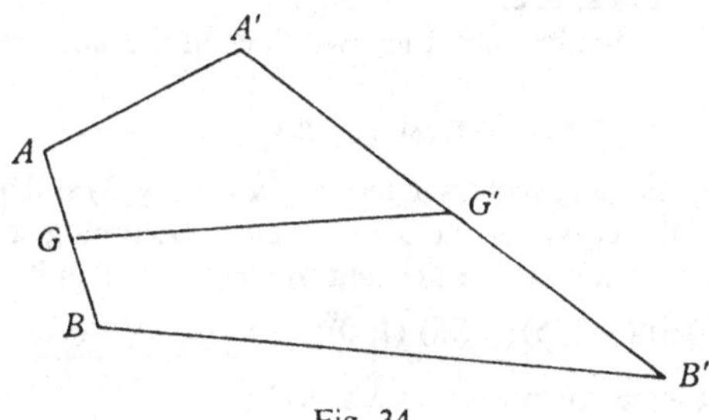

Fig. 34

4. Prove by vectors that the line joining the mid-points of AB and AC is parallel to BC and equal to half BC in length.

Take A as origin and let $\mathbf{AB} = \mathbf{b}$, etc. Then $\mathbf{h} = \frac{1}{2}\mathbf{b}$ and $\mathbf{k} = \frac{1}{2}\mathbf{c}$. Thus $\mathbf{HK} = \mathbf{k}-\mathbf{h} = \frac{1}{2}(\mathbf{c}-\mathbf{b}) = \frac{1}{2}\mathbf{BC}$, and the result follows.

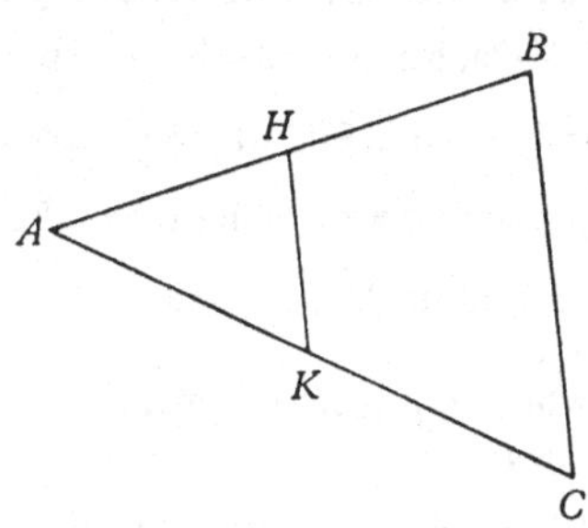

Fig. 35

5. A triangle ABC suffers two displacements in its plane: (i) a reflection about a point O to a position UVW, so that O is the mid-point of each of AU, BV, CW; (ii) a parallel displacement to a position XYZ, so that AX, BY, CZ are equal and parallel. The centres of the parallelograms $BYZC$, $CZXA$, $AXYB$ are L, M, N, respectively. Establish the existence of a point K for which L, M, N are mid-points of KU, KV, KW, respectively.

(Cambridge Open Scholarship.)

Take O as origin and let $\mathbf{OA} = \mathbf{a}$, etc. Then $\mathbf{u} = -\mathbf{a}$, $\mathbf{v} = -\mathbf{b}$, $\mathbf{w} = -\mathbf{c}$. Let $\mathbf{AX} = \mathbf{BY} = \mathbf{CZ} = \mathbf{d}$, so that $\mathbf{x} = \mathbf{a}+\mathbf{d}$, $\mathbf{y} = \mathbf{b}+\mathbf{d}$, $\mathbf{z} = \mathbf{c}+\mathbf{d}$. Then $\mathbf{l} = \frac{1}{2}(\mathbf{b}+\mathbf{z}) = \frac{1}{2}(\mathbf{b}+\mathbf{c}+\mathbf{d})$, since L is the mid-point of BZ, and $\mathbf{m} = \frac{1}{2}(\mathbf{c}+\mathbf{a}+\mathbf{d})$, $\mathbf{n} = \frac{1}{2}(\mathbf{a}+\mathbf{b}+\mathbf{d})$ similarly.

To establish the existence of K, we need to find the points that divide UL, VM, WN externally in the ratio $-2:1$ and to show that these points coincide in K. The point that divides UL in the ratio $-2:1$ is

$$\frac{-2\mathbf{l}+\mathbf{u}}{-2+1} = 2\mathbf{l}-\mathbf{u}$$
$$= \mathbf{b}+\mathbf{c}+\mathbf{d}-(-\mathbf{a})$$
$$= \mathbf{a}+\mathbf{b}+\mathbf{c}+\mathbf{d}$$

and is symmetrical in **a**, **b**, **c**.

Hence this point divides the other two lines in the same ratio.

Exercises 8A

1. Express the following vectors in terms of $\mathbf{a} = (\frac{1}{2}, \frac{1}{2}\sqrt{3})$ and $\mathbf{b} = (-\frac{1}{2}\sqrt{3}, \frac{1}{2})$ and verify that if (x_1, y_1) is the given vector, becoming $x_2\mathbf{a}+y_2\mathbf{b}$, then $x_1^2+y_1^2 = x_2^2+y_2^2$. (**a** and **b** are at right angles and of unit length):

(i) (2, 3); (ii) (−2, 5); (iii) (1, 0); (iv) (0, 1); (v) (−3, −7).

2. Show that the vector product $\mathbf{a} \wedge \mathbf{a} = \mathbf{0}$.

3. If **x**, **y** are vectors in three dimensions prove:

(i) $|\mathbf{x}+\mathbf{y}| \leqslant |\mathbf{x}| + |\mathbf{y}|$;
(ii) $|\mathbf{x}+\mathbf{y}| \geqslant |\mathbf{x}| - |\mathbf{y}|$;
(iii) $|\mathbf{x}-\mathbf{y}| \geqslant |\mathbf{x}| - |\mathbf{y}|$;

4. Draw vectors **a** and **b**. Construct the following vectors:

$$-2\mathbf{a},\ \mathbf{a}+2\mathbf{b},\ -\mathbf{a}-\mathbf{b},\ \tfrac{1}{2}\mathbf{a}-\tfrac{2}{3}\mathbf{b}.$$

5. If $\mathbf{OA} = \mathbf{a}$, $\mathbf{OB} = \mathbf{b}$, what is $\mathbf{OC}$ where C bisects AB?

6. If $\mathbf{OA} = \mathbf{a}$, etc., prove that $\mathbf{a}+\mathbf{c} = \mathbf{b}+\mathbf{d} \Leftrightarrow ABCD$ is a parallelogram.

7. If **a** and **b** are fixed, find the locus of **x** if:

(i) $|\mathbf{x}-\mathbf{a}| = |\mathbf{x}-\mathbf{b}|$;
(ii) $|\mathbf{x}-\mathbf{a}| = \lambda$, where λ is a fixed scalar;
(iii) $|\mathbf{x}-\mathbf{a}| = 2|\mathbf{x}-\mathbf{b}|$.

8. If A', B', C' are the mid-points of BC, CA, AB prove that the triangles ABC, $A'B'C'$ have the same centroid.

9. If A', B', C' divide BC, CA, AB respectively in the ratio $k_1:k_2$ prove that the triangles ABC, $A'B'C'$ have the same centroid.

10. If P, Q, R, S are the centroids of the faces BCD, CDA, DAB, ABC of a tetrahedron $ABCD$, prove that $ABCD$ and $PQRS$ have the same centroid.

11. If $\mathbf{OA} = \alpha$, $\mathbf{OB} = \beta$ where α and β are complex numbers, prove that

$$|\alpha|^2 + |\beta|^2 = \tfrac{1}{2}|\alpha+\beta|^2 + \tfrac{1}{2}|\alpha-\beta|^2.$$

12. If A, B, C, D are (3, 0), (2, 5), (−1, −3), (−5, 7) respectively what are **AB**, **CD**, **AC**, **DB**, **BD**?

13. Prove that, if O is the circumcentre and H the orthocentre of the triangle ABC, then

(i) $\mathbf{OA}+\mathbf{OB}+\mathbf{OC} = \mathbf{OH}$;

(ii) $\mathbf{HA}+\mathbf{HB}+\mathbf{HC} = 2\mathbf{HO}$.

14. Prove that if P, Q, R, S bisect AB, BC, CD, DA then $PQRS$ is a parallelogram.

15. Prove by vectors that the diagonals of a parallelogram bisect each other.

16. If $\sum_{k=1}^{n} |z_k| = \left|\sum_{k=1}^{n} z_k\right|$ show that there is a complex number z_0 such that, for $1 \leqslant k \leqslant n$, z_k/z_0 is real and non-negative.

(Cambridge Open Scholarship.)

17. Prove that $\lambda\mathbf{a} = \mathbf{0} \Rightarrow \lambda = 0$ or $\mathbf{a} = \mathbf{0}$.

18. List all the two-dimensional vectors whose components are residues modulo 2 and give the sum of each pair.

19. In the three-dimensional vectors with coefficients in the set of residues modulo 5, if $\mathbf{a} = (1, 3, 2)$, $\mathbf{b} = (2, 1, 0)$, $\mathbf{c} = (4, 0, 1)$, find

$$\mathbf{a}+2\mathbf{b},\ \mathbf{a}+\mathbf{b}+\mathbf{c},\ \mathbf{c}-2\mathbf{b}-\mathbf{a}.$$

Exercises 8B

1. If $\mathbf{a} = (a_1, a_2)$ and $\mathbf{b} = (b_1, b_2)$ are represented geometrically by **OA** and **OB** and if $\angle AOB = \theta$ prove by the cosine rule that

$$a_1b_1+a_2b_2 = OA.OB\cos\theta.$$

Extend this to three dimensions where $\mathbf{a} = (a_1, a_2, a_3)$ and $\mathbf{b} = (b_1, b_2, b_3)$ and we consider the inner product $a_1b_1+a_2b_2+a_3b_3$.

2. If $\mathbf{a} = (1, 1, 0)$, $\mathbf{b} = (1, 0, 0)$, $\mathbf{c} = (0, 1, 0)$ express the following vectors in terms of **a**, **b**, **c** wherever possible. If it is not possible say so. In the possible cases give the general expression:

(i) $(3, 5, 0)$; (ii) $(0, 0, 1)$; (iii) $(1, 1, 0)$; (iv) $(-2, 0, 1)$.

3. Prove that if z_1, z_2, z_3, z_4 are complex numbers and

$$\frac{(z_1-z_3)}{(z_3-z_2)}\bigg/\frac{(z_1-z_4)}{(z_4-z_2)}$$

is real then Z_1, Z_2, Z_3, Z_4 are either collinear or concyclic. (*Note.* The condition is that the cross-ratio (z_1z_2, z_3z_4) is real.)

4. If $\mathbf{OA} = a$, etc., where $a, \ldots$ are complex numbers, prove that if the triangles ABC, PQR are directly similar (i.e. each can be transformed into the other by an enlargement plus a rotation and translation in the plane, without turning over) then $(a-b)/(c-b) = (p-q)/(r-q)$ and conversely.

Show that this condition is

$$\begin{vmatrix} 1 & 1 & 1 \\ a & b & c \\ p & q & r \end{vmatrix} = 0.$$

5. If G is the centroid of masses m_i at the coplanar points A_i ($i = 1 \dots n$) and P is any point in the plane prove that

$$\sum_i m_i PA_i^2 = \sum_i m_i GA_i^2 + (\sum_i m_i) PG^2.$$

(The parallel axis theorem for moments of inertia.)

6. Prove that if G is the centroid of $A_1, \dots, A_n$ and H is the centroid of $B_1, \dots, B_n$ then $\sum_{i=1}^{n} \mathbf{A}_i\mathbf{B}_i = n\mathbf{GH}$.

7. If $\mathbf{OA} = \mathbf{a}$, $\mathbf{OB} = \mathbf{b}$, show that the equation of the line AB is $\mathbf{r} = t\mathbf{b} + (1-t)\mathbf{a}$, where $\mathbf{r} = \mathbf{OR}$, R being any point on the line, and t is a scalar parameter.

8. Prove by vectors Desargues' theorem, that if AA', BB', CC' are concurrent, then the meets of BC and $B'C'$, CA and $C'A'$. AB and $A'B'$ are collinear, and conversely.

9. The complex numbers a, b, c are represented in the Argand diagram by the points A, B, C. Show that ABC is an equilateral triangle if and only if a, b, c are not all equal and $a^2+b^2+c^2-bc-ca-ab = 0$. Three equilateral triangles $X'YZ$, $Y'ZX$, $Z'XY$ are drawn outwards from the sides of the triangle XYZ. Show that the triangles XYZ, $X'Y'Z'$ have a common centre of gravity. (Cambridge Open Scholarship.)

10. On the sides of a triangle $Z_1Z_2Z_3$ are constructed isosceles triangles $Z_2Z_3W_1$, $Z_3Z_1W_2$, $Z_1Z_2W_3$ lying outside the triangle $Z_1Z_2Z_3$. The angles at W_1, W_2, W_3 are all $\frac{2}{3}\pi$. By assigning complex numbers z_1, z_2, z_3 to Z_1, Z_2, Z_3 and calculating the numbers representing W_1, W_2, W_3, or otherwise, prove that $W_1W_2W_3$ is equilateral.

(Cambridge Open Scholarship.)

9

FUNCTIONS AND MAPPINGS

9.1. The idea of a function

The idea of a function is fundamental to any advanced or fairly advanced mathematical work. If we have a variable y whose value is uniquely determined by the value of another variable x, we say that y is a function of x and write $y = f(x)$. x is called the *independent variable*, and y the *dependent variable.*

Practical examples of functions are:

(i) y is the distance gone by a body in time x;

(ii) y is the gravitational force exerted on a body at distance x from the earth;

(iii) y is the side of a square whose area is x;

(iv) y is the sum of the digits of x, where x is expressed as a decimal;

(v) y is the temperature at noon on day x;

(vi) y is the time in which a body will travel a distance x;

(vii) y is the amount to which a certain sum of money invested at compound interest has grown after a time x years.

Many functions can be expressed symbolically in terms of the basic mathematical functions, e.g. as polynomials or rational functions or in terms of the circular, exponential or logarithmic functions. Thus in (i) if the body starts from rest and is moving under gravity, $y = \frac{1}{2}gx^2$, in (ii) $y = k/x^2$, while in (vii) $y = Ae^{Bx}$ for constants A and B. Not all functions are expressible thus: there are no simple expressions for (iv) and (v).

y may not be defined for all values of x. Thus in (vi) where the body is falling from rest under gravity, $y = \sqrt{(2x/g)}$ and is not defined (in terms of real numbers) for any negative value of x. Another cause of failure arises in (iv), where if x is not a finite decimal, y is undefined (being infinite).

x and y may range over various sets. We usually tend to think of them as real numbers, but this is by no means always so even in quite elementary work. x is often restricted to integers (in examples (v) and (vii), although non-integral values of x may be interpolated in (vii)), sometimes it is effectively restricted to positive real numbers, as in (iii) and (vi) as discussed above, while it often helps to include complex values as well (e.g. in solving quadratic equations, to find where $f(x) = 0$). y may similarly be restricted or extended to sets other than the reals (positive integers in (iv), complex numbers in the general case of (iii)).

The set over which x is considered to range is called the *object space* of the function; that over which y ranges is the *image space*.

It is very often useful to represent a function by a graph in the usual way. This is particularly valuable when both the object and image spaces are the real numbers, but it can also be done for any subsets of these (e.g. the integers). When x and y may be complex a graph would need four dimensions in which to be drawn.

9.2. Transformations in geometry

An idea similar to that of a function is often used in geometry. Certain transformations of geometrical figures often lead to other configurations that can be dealt with more easily. We give a few examples, noting their similarities to functions and emphasising the sets that form the object and image spaces. In all these cases the image will in general be uniquely defined.

Inversion

In a plane we invert with respect to a point O and a circle centre O radius k as follows:

We define the inverse point P' of a point P as that point on OP such that $OP \, . \, OP' = k^2$. This definition transforms each point in the plane into a unique point, except that O itself is transformed into no finite point, but into the whole 'line at infinity', which consists of all the points 'at infinity' in any direction. The object and the image spaces here are all points of the plane.

If P describes a curve Γ, P' will describe another curve Γ', the *inverse curve* to Γ. This gives a transformation between curves in the plane. The importance of the idea arises from the fact that if Γ is a straight line or circle then so is Γ', in particular if Γ is a circle through O then Γ' is a straight line, and conversely. Thus inversion is a transformation of the set of all straight lines and circles into the same set, but *not* of the set of circles or that of straight lines into itself.

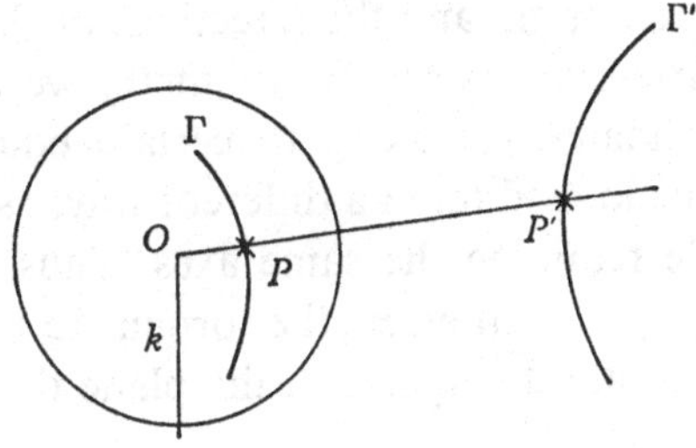

Fig. 36

In three-dimensional inversion we replace the original circle of inversion by a sphere centre O, and as before define the inverse of a point, then of a curve and similarly of a surface. Spheres and planes invert into spheres and planes, spheres through O becoming planes.

Reciprocation

In reciprocation in a plane we transform points in the plane into their polars with respect to a fixed conic in the plane, and lines into their poles with respect to the same conic. Thus if the object space is the set of points the image space is the set of lines, and conversely. If we reciprocate the points of a curve the lines we obtain will envelope a second curve, defined as the reciprocal of the first. A conic reciprocates into a conic. Every point reciprocates into a unique line if we admit the inclusion of the line at infinity.

Reciprocation is a special case of the principle of duality, for the details of which the reader should see any standard textbook on projective geometry.

In three dimensions we reciprocate by transforming points into their polar planes with respect to a fixed quadric and planes into their poles with respect to the same quadric. A line of points reciprocates into the line of intersection of the polar planes of the points. The points of a surface become planes all of which touch the reciprocal surface, while a curve reciprocates into another curve.

Change of axes in Cartesian geometry

If we change the origin and the directions of the axes in two-dimensional Cartesian co-ordinate geometry we need to transform the co-ordinates according to certain equations. These equations may be looked on in a different way, as transforming the figures while retaining the same axes. Thus the equations $X = x+h$, $Y = y+k$ transfer the origin to $(-h, -k)$, or alternatively translate the figures in the plane through a vector (h, k). The equations

$$X = x \cos \theta + y \sin \theta, \quad Y = -x \sin \theta + y \cos \theta$$

either rotate the axes through an angle θ anticlockwise or keep the axes fixed and rotate the figures through θ clockwise.

When we look upon the equations as transforming the geometrical figures we obtain a transformation in which every figure becomes a congruent one, lengths and angles being unaltered. The line at infinity transforms into itself.

Projection in Cartesian geometry

We take two planes in space, π and π', and a point O not in either. Any point P in π is projected into its image P' in π' by joining OP and letting it meet π' in P'. We obtain a transformation of the points of π into those of π', and conversely of those in π' to those in π, by the reverse process.

Distances and angles are changed by this transformation, in general, and the line at infinity transforms into a finite line in π'. The fundamental property which is unchanged is the cross-ratio of four collinear points or concurrent lines. (This is proved in any standard work on the subject.) Also, lines project into lines and conics into conics, but circles will in general become conics which are not circles. With any origins and axes in π and π' the equations connecting the co-ordinates will be of the form

$$\frac{X}{l_1x+m_1y+n_1} = \frac{Y}{l_2x+m_2y+n_2} = \frac{1}{l_3x+m_3y+n_3},$$

where

$$\Delta = \begin{vmatrix} l_1 & m_1 & n_1 \\ l_2 & m_2 & n_2 \\ l_3 & m_3 & n_3 \end{vmatrix} \neq 0.$$

Linear transformation in projective geometry

In projective geometry we use homogeneous co-ordinates, and may transform them linearly, the most general two-dimensional such transformation being

$$X = l_1x+m_1y+n_1z,$$
$$Y = l_2x+m_2y+n_2z,$$
$$Z = l_3x+m_3y+n_3z,$$

where

$$\Delta = \begin{vmatrix} l_1 & m_1 & n_1 \\ l_2 & m_2 & n_2 \\ l_3 & m_3 & n_3 \end{vmatrix} \neq 0.$$

As in the case of Cartesian change of axes this may be thought of in two ways. The first is as a change of axes—in this case a change of triangle of reference and unit point. The other way is as a transformation of the geometry keeping the same triangle of reference and unit point. In this case points and lines transform into points and lines (always unique) and conics into conics. Cross-ratio is unaltered.

Change of Cartesian axes and projection are special cases of this general linear transformation, though projection in fact includes all such equations, and gives the name 'projective' to this type of geometry, projective geometry being the study of properties that are unaltered by a general linear transformation.

In the above transformation, if $\Delta = 0$ then although each point transforms into a unique point (X, Y, Z), two different points may transform into the same point. The new triangle of reference is in fact degenerate, consisting of three concurrent lines.

9.3. Mappings

In the preceding section we saw how an idea similar to that of functions may be used in geometry. We may extend the process still further, to apply to any sets. Thus if to every element a of a set A there is associated a unique element b in another set B we say that we have a *mapping* from the set A into the set B.

The sets A and B may be the same or different. Note that *every* element of A must give rise to a *unique* element of B. Thus the function which sends x into $\pm\sqrt{x}$ is not a mapping of any set except that consisting of zero alone, since $\pm\sqrt{x}$ has two values. If we restrict ourselves to real x and interpret $\sqrt{x}$ as being positive or zero we still do not have a mapping of the set of reals into itself, since if x is negative $\sqrt{x}$ does not exist in terms of real numbers. We *do*, however, have a mapping of the set of positive or zero real numbers into itself, or indeed of the positive or zero real numbers into the complete set of reals, since there is no need for all the elements of the second set to arise from some element of the first. Two elements of A may map into the *same* element of B.

Nomenclature

The set A is called the *object space* (or object set) or *domain*, B is the *image space* (or image set) or *codomain*, a is the *object* whose *image* is b. We say that we have a mapping of A *into* B.

Notation

Functions are usually denoted thus: $f(x)$. For mappings we will usually, though by no means always, use small Greek letters, θ, ϕ, ψ, etc. If the mapping θ is between the object space A and image space B we write $\theta: A \to B$ (a mapping θ of A into B).

There are two schools of thought as to the better way of denoting the image of an object a under a mapping θ. The first, extending the notation for a function, writes $\theta(a)$, while the second writes $a\theta$. The first way may seem the more natural at first sight, but the second is in fact easier to use in advanced work. The main reason for this can be seen when we consider products of mappings in §9.6. Some writers even use a^θ, while it is occasionally convenient to have a mixture of the first two notations. Feeling on the subject runs high, but the author believes firmly in the notation $a\theta$, and this will be used here.

The importance of the idea of a mapping can hardly be overestimated. The reader should already be convinced of the fundamental role of a function in more elementary mathematics, and when we are dealing with sets other than numbers mappings take over much of its importance. They give us a method of going from one set to another, forming a connection, a sort of bridge, between the sets.

Examples of mappings

We give here a selection of the many examples. The reader should easily be able to construct others.

(1) *Functions of real or complex variables*

Some of these have been mentioned in §9.1. Functions of two variables form a mapping of a plane into a line, and need three dimensions for their graphical representation if the variables are all real, or six dimensions if complex.

(2) *Practical examples*

There are many obvious ones, though they need to be translated into mathematical terms before we can work with them. For example, if we have a set of boys, each of whom is marked in, say, eight subjects, we obtain a mapping of the set of boys into the set of eight-dimensional vectors. Here $x\theta = (a_x, b_x, ..., h_x)$ where the boy x obtains marks $a_x, b_x, ..., h_x$ in the subjects taken in some order.

We may obtain a mapping of the days of a year into the days of the week by taking $x\theta$ to be the day on which x falls in any given year. We obtain a separate mapping for each year, with eventual repetitions.

We may define a mapping of all human beings, alive or dead, into itself by defining the image of a person to be that person's father. Note that if the image were defined as a parent this would not be mapping, since the definition is not unique. Nor would the definition of image as eldest child be a mapping, since this would not always exist. But the definition of eldest child *is* a mapping between the set of all parents and the set of all humans.

(3) *Sequences*

A sequence is a set $a_1, a_2, \ldots, a_n, \ldots$, where an a_i is defined for every positive integer i. If the a_i's belong to a set B then a sequence may be thought of as a mapping of the set Z of positive integers into B, i.e. as a mapping $\theta: Z \to B$. B will often be the reals R.

(4) *Mappings of polynomials*

A mapping of the set of all polynomials into the set Z of positive or zero integers is given by $\theta: P = d(P)$ where P is a polynomial and $d(P)$ is the degree of P.

$\theta: P \to dP/dx$ gives a mapping of the set of polynomials onto itself; under this mapping the subset of polynomials of degree n has as image the subset of those with degree $(n-1)$.

$\theta: P \to \int P.dx$ does not give a mapping since it is not unique, but $\theta: P \to \int_0^x P(t).dt$ *is* a mapping.

(5) *Mappings involving residues*

$\theta: x \to$ the residue class of x modulo n is a mapping of the integers into the set of residues modulo n. Note that under this mapping $(x+y)\theta = x\theta + y\theta$ and $(xy)\theta = (x\theta)(y\theta)$, where the addition and multiplication on the right-hand side is taken modulo n. (A mapping like this, which preserves the algebraic processes, is called a *homomorphism*.)

Since all numbers with the same residue modulo 6 will have the same residue modulo 2 we may define a mapping of the set R_6 of residues modulo 6 into the set R_2 by defining $x\theta$ to be the residue modulo 2 of any member of x (considered as an element of R_6). The same may be done for any two moduli mn and m, but not in other cases.

Residue classes are an example of equivalence classes between the integers (§6.2) and we may define a mapping generally from any set to the set of equivalence classes under some equivalence relation, by defining $a\theta$ to be that class which contains a. Note that we cannot reverse the process, since there would in general be more than one element in each class.

(6) *Permutations*

Suppose we have a set A containing a finite number n of elements which we label $1, 2, \ldots, n$. A mapping of A into itself which sends each element into a different one is called a *permutation* of the elements of A. Permutations are studied in §12.6.

(7) *Mappings involving subsets*

We may have a mapping from a set into a subset of it. A good example is that of orthogonal projection. In the set of points in two dimensions we define the projection (orthogonal) of a point P on to a fixed line l in the plane by drawing the perpendicular PP' through P to l. P' is the orthogonal projection of P. If the line l is taken as Ox and any line at right angles to it as Oy, then the mapping sends the point (or vector) $P(x, y)$ into $P'(x, 0)$. This may also be considered as a mapping of the plane into a line, or as a mapping of the set of two-dimensional vectors into the one-dimensional vectors, i.e. the scalars.

Similarly, we may project three-dimensional space orthogonally on to either a plane or a line. To project on to a plane we take the foot of the perpendicular to the plane, thus in terms of co-ordinates $(x, y, z) \rightarrow (x, y, 0)$ is the projection on to the plane $z = 0$. To project on to a line we take the foot of the perpendicular to the *line*: thus $(x, y, z) \rightarrow (x, 0, 0)$ is the projection on to the x-axis.

The simplest mapping of a subset into a set is that known as an *injection*. The injection which maps the subset B into the set A which contains B is defined by $\theta: b \rightarrow b$, where the left-hand b, the object, is considered as an element of B, while the right-hand b, the image, is considered as an element of A (which contains B and therefore contains all $b \in B$).

As an example of a mapping from a subset to a set which is not an injection we consider θ: the even integers $\rightarrow$ the complete set of integers, defined by $(2n)\theta = n$.

9.4. 1-1 correspondence

Inverse functions and mappings

If we have a function $f(x) = y$ we often find it useful to consider the function that gives x when operating on y, i.e. such that $g(y) = x$. For example, if $y = x^2$ then $x = \sqrt{y}$, if $y = e^x$ then $x = \log y$, if $y = \sin x$ then $x = \sin^{-1}y$. (To give a *function* we must choose just one of the possible values for x: thus $\sin^{-1}y$ is taken to be in the range $-\pi/2 \leqslant \sin^{-1}y \leqslant \pi/2$.) Such functions are called *inverse functions*.

In the same way, inverse mappings are important. The inverse of a mapping $\theta: A \to B$ would be a mapping $\phi: B \to A$ such that if $a\theta = b$ then $b\phi = a$. Inverse mappings do not, however, always exist. Since under a mapping each object must have one and only one image, the necessary and sufficient condition for an inverse of θ to exist is that every element of B must be the image of one and only one element of A, under θ. We need in effect two conditions on the mapping θ.

(1) Each element of B must be the image of some $a \in A$, i.e. the images of all elements of A must together cover the whole of B. A mapping that has this property is called a mapping of A *onto B*. We distinguish carefully here between the words into and onto. A general mapping of A will be *into B*, but to say that it is *onto B* means that it satisfies the condition that the images cover the whole of B. (2) The second condition for θ to have an inverse is that each element of B must be the image of only one element of A, i.e. if $a\theta = a'\theta$ then $a = a'$. Such a mapping is said to be 1-1 ('one-one', or 'one to one').

Mappings may be onto and not 1-1, or 1-1 but not onto (see examples below). If θ is both onto and 1-1 then an inverse mapping, in the sense defined above, exists—each element of B arises from one and only one element of A. The inverse mapping is denoted by θ^{-1}. The reason for this notation is that an inverse mapping behaves rather like the reciprocal or inverse of a number (this will be seen more clearly in §9.6). The notation has already been met with in the inverse circular and hyperbolic functions.

The definition of θ^{-1}, for a mapping $\theta: A \to B$ which is both onto and 1-1, is then that θ^{-1} is a mapping of B into A such that if $a\theta = b$ then $b\theta^{-1} = a$; the image of b under θ^{-1} is that element which goes to b under θ, object and image are interchanged. Notice that θ^{-1} is both onto and 1-1, since θ acts on the whole of A, uniquely, and that the inverse of θ^{-1} is θ.

The set of all the images of a subset A', of A, under θ, is a subset of B and is called the image of A', being denoted by $A'\theta$. The set of all the images of elements of A is called the image of the mapping (*not* the image space, which is B): it is a subset of B and is denoted by $A\theta$. $A\theta = B$ is the condition that θ is onto.

Correspondences

A correspondence between sets A and B is an operation that associates with each element of A one or more elements of B and vice versa (in some cases there may be no elements corresponding to certain ones). A mapping is a many-one correspondence, several (many) elements of A may correspond to one of B, but an element of A corresponds to just one element of B. We could have one-many correspondences (e.g. a mapping of B into A), or even many-many. (An example of the latter is that of parent and child where both A and B are the set of all human beings, living or dead.)

A mapping that is onto and 1-1 is often known as a 1-1 correspondence. This is by far the most important type of correspondence, and gives a pairing off of the elements of A against those of B, each corresponding to a unique element of the other set. Note that a mapping that is 1-1 but not onto does not give a 1-1 correspondence between A and the whole of B, but *does* do so between A and a subset of B.

Examples

Mappings which are onto but not 1-1. The mapping of the reals into the positive or zero reals defined by $x\theta = x^2$. But the same mapping operating on the positive or zero reals only is both onto and 1-1 and has the inverse $x \to \sqrt{x}$.

The mapping of polynomials into polynomials defined by $P\theta = dP/dx$, since if P and P' differ by a constant their derivatives are the same. If we ignore the constant differences *in the object set only* then we have a 1-1 mapping: its inverse is the integral.

The mapping of an integer x into its residue modulo n.

Orthogonal projections as in example 7 of §9.3.

Mappings which are 1-1 *but not onto.* The mapping

$$P \to \int_0^x P(t)dt$$

of the set of polynomials into itself, since no P will have an image including a constant. But as above we can get over this difficulty by ignoring the constants in the image set.

Injections as defined in example 7 of §9.3.

By extending the image set of any 1-1 correspondence we can obtain a mapping that is not onto, but still 1-1. Conversely, given any 1-1 mapping $\theta: A \to B$ we may make it into an onto mapping by restricting the image space to consist of $A\theta$ only.

Mappings which are both onto and 1-1: 1-1 *correspondences.* The inversion mapping of all points in a plane is 1-1 and onto if we exclude the centre of inversion; it gives a 1-1 correspondence between the points, and in this case the inverse is the same mapping. It also gives a 1-1 correspondence between the set of all lines and circles in the plane and itself.

Reciprocation gives a 1-1 correspondence between the set of all points in the plane and the set of all lines, if we include the line at infinity. The inverse mapping is between the set of lines and that of points.

Projective geometry. The reader has probably met the term '1-1 correspondence' in connection with homographies and cross-ratio. If the points on two lines or curves are parametrised and the parameters are connected by an expression of the form $a\lambda\lambda' + b\lambda + c\lambda' + d = 0$, where a, b, c, d are constants, λ and λ' are the parameters and $ad \neq bc$, then there is a 1-1 correspondence between the two sets of points. Such a correspondence is called a *homography* or *projectivity* and the basic theorem is that the cross-ratio of four λ's is equal to the cross-ratio of the corresponding λ''s.

The extension of homographies to the points of a plane is given by a linear transformation as in §9.2, and this is another example of a 1-1 correspondence between the set of points of the plane and itself. The correspondence here is not only 1-1 but also algebraic and rational.

Finite sets. A permutation, or re-arrangement, of the numbers $1, 2, \ldots, n$ is a 1-1 correspondence. Conversely, if there is a 1-1 correspondence between the elements of two *finite* sets A and B then A and B must contain the same number of elements. This leads to one of the basic definitions of positive integers, the *cardinal numbers*. Two sets are said to have the same cardinal number if their elements can be put into 1-1 correspondence. We then obtain the finite cardinals $1, 2, \ldots$, the number n being thought of as the property that certain sets with the same cardinal have in common (in line with the idea of 2 being the property common to all pairs of objects or elements).

Infinite sets. Infinite sets are still said to have the same cardinal number if they can be put in 1-1 correspondence. This leads to the so-called 'transfinite numbers' which are the cardinals of infinite sets. The smallest of these is the 'denumerable infinity', often called $\aleph_0$ (aleph-nought). A set has cardinal $\aleph_0$ if it can be put in 1-1 correspondence with the set of positive integers. Such a set is called 'denumerable'. We showed in theorem 5.1.2 that the set of rationals is denumerable, while in theorem 5.2.2 we proved that the set of reals is not.

9.5. Isomorphisms

We sometimes have a 1-1 correspondence between two sets which preserves all the algebraic processes with which we are concerned in a given piece of work. As an example we recall the definition of complex numbers as pairs (x, y) of real numbers. If we restrict ourselves to those complex numbers of the form $(a, 0)$ we may put these in 1-1 correspondence with the reals by $(a, 0) \leftrightarrow a$. Then we found that $(a, 0)+(b, 0) = (a+b, 0)$ and $(a, 0).(b, 0) = (ab, 0)$ and so the correspondence preserves addition and multiplication, and of course subtraction and division also. Provided we restrict ourselves to properties such as these that are not altered by the correspondence it is immaterial whether we deal with the set $(a, 0)$ or the set of reals, and we do in fact identify them and talk about the 'real number' $(a, 0)$ and write it in the form a. Two such sets are known as *isomorphic* sets, and the correspondence is called an *isomorphism*. Note that they are, strictly speaking, different sets (a is *not* the *same* as $(a, 0)$, one being a real number and the other a pair of reals), but so long as we keep to properties that are unaltered by the correspondence we may treat them in exactly the same way. An example of a property that *is* altered is, in this case, multiplication by (x, y): this is not defined for a real number, but is for the pair $(a, 0)$.

A good example of isomorphism occurs with vectors. A vector may be thought of as a displacement, a point, a set of co-ordinates or, in the two-dimensional case, as a complex number. Thus in the case of two-dimensional vectors we have the following isomorphic sets:

(1) the set of displacements in a plane, equal and parallel ones being considered to be the same, with addition defined by the triangle rule,

(2) the set of points in a plane, addition being defined in terms of co-ordinates,

(3) the set of ordered pairs (x, y) of real numbers, addition being defined by $(x_1, y_1)+(x_2, y_2) = (x_1+x_2, y_1+y_2)$,

(4) the set of numbers $x+iy$ where $i^2 = -1$, addition defined as usual (note that our definition of complex numbers was in fact in the form of (3)).

These sets are isomorphic under the correspondence

$$\mathbf{OP} \leftrightarrow P \leftrightarrow (x, y) \leftrightarrow x+iy$$

for all processes derived from addition, and with the usual definitions for multiplication by a scalar. We cannot use the isomorphism for multiplicative properties of two vectors without giving suitable definitions of this: this can of course be done but not in any obvious way in cases (1), (2) or (3). In fact the multiplicative structure in the first three cases is derived from that of (4) and does not carry over into higher dimensions.

Another set which is isomorphic to vectors in n-dimensions is that of all polynomials of degree $< n$, for addition and multiplication by a scalar, but not for general multiplication, although this may be defined for polynomials provided there is no restriction of degree.

In geometry the sets of points (in terms of homogeneous coordinates) with respect to two different triangles of reference are isomorphic, the correspondence being given as a linear transformation. The isomorphism applies to all projective properties. In Euclidean geometry, under rotations and translations, a similar isomorphism holds and applies to Euclidean properties such as distance and angle.

9.6. Products of mappings

Suppose we have a mapping $\theta: A \to B$ and a mapping $\phi: B \to C$. If $a\theta = b$ and $b\phi = c$ then we may consider the mapping that sends a into c: this is a mapping of A into C known as the *product* of θ and ϕ and denoted by $\theta\phi$. Note that it is a properly defined mapping, since $a\theta$ is unique for any $a \in A$ and, since $a\theta$ is in B, its image is uniquely defined in C.

The image of a under θ is $a\theta$, and the image of this under ϕ is $(a\theta)\phi$, which may be written merely as $a\theta\phi$. We see here the reason for our choice of this notation, for with the notation $\theta(a)$ we would have to write the product as $\phi(\theta(a))$, or $\phi\theta(a)$ and here the *second* mapping must be written first, while with our notation the first mapping appears first, and this is less confusing when we have a string of mappings.

We may of course form the product of any number of mappings, provided always that the image space of each is the same as the object space of the next one (or is a subset of it). It is often convenient to make a diagrammatic representation as in figure 37.

$$A \xrightarrow{\theta} B \xrightarrow{\phi} C \xrightarrow{\psi} D$$

Fig. 37

The product of two mappings is analogous to the idea of a function of a function; thus if $y = f(x)$ and $z = g(y)$ we have $z = g(f(x))$.

If $\theta: A \to B$ is onto and 1-1, so that θ^{-1} exists, we have $a\theta\theta^{-1} = a$ and $b\theta^{-1}\theta = b$. The mapping that sends each element into itself is called the identity mapping I, or sometimes I_A if it acts on the space A. With this notation $\theta\theta^{-1} = \theta^{-1}\theta = I$, or more strictly, $\theta\theta^{-1} = I_A$ and $\theta^{-1}\theta = I_B$. Note the analogy with ordinary numbers where x multiplied by its reciprocal equals the unity 1.

Theorem 9.6.1. *If θ and ϕ are both onto and* 1-1, *so is $\theta\phi$, and* $(\theta\phi)^{-1} = \phi^{-1}\theta^{-1}$.

If $\theta: A \to B$ and $\phi: B \to C$ we have that $A\theta = B$ and $B\phi = C$ so that $A\theta\phi = C$, so that $\theta\phi$ is onto. Suppose $\theta\phi$ is not 1-1, so that $a\theta\phi = a'\theta\phi$ for some a and a'. Then since ϕ is 1-1 we must have $a\theta = a'\theta$ and since θ is 1-1 this implies that $a = a'$. Hence $\theta\phi$ is 1-1.

ϕ^{-1} is a mapping of C onto B and θ^{-1} is one of B onto A, and so $\phi^{-1}\theta^{-1}$ maps C onto A. Further, if

$$\begin{aligned} a\theta\phi = c, \quad c\phi^{-1}\theta^{-1} &= a\theta\phi\phi^{-1}\theta^{-1} \\ &= a\theta(\phi\phi^{-1})\theta^{-1} \\ &= a\theta I\theta^{-1} \\ &= a\theta\theta^{-1} \\ &= a. \end{aligned}$$

Hence $\phi^{-1}\theta^{-1} = (\theta\phi)^{-1}$.

We show this diagrammatically in figure 38.

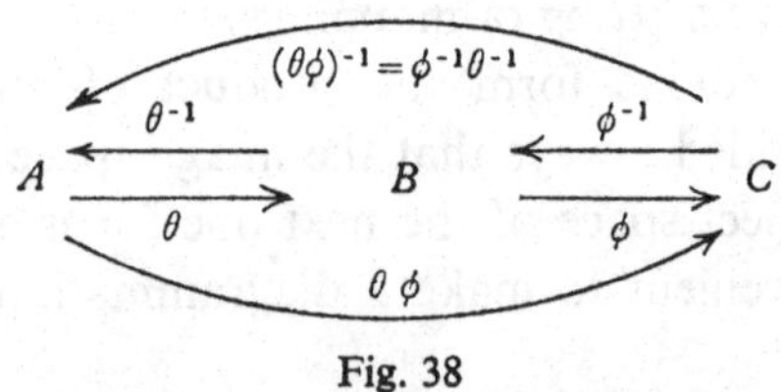

Fig. 38

Involutions

If θ is a 1-1 correspondence between elements of A and itself such that $\theta = \theta^{-1}$ then θ is said to be an involution. The condition is equivalent to $\theta^2 = I$, since if $\theta = \theta^{-1}$ then

$$\theta^2 = \theta\theta^{-1} = I,$$

while if $\theta^2 = I$ we have $\theta^2\theta^{-1} = \theta^{-1}$ and so $\theta(\theta\theta^{-1}) = \theta^{-1}$, i.e. $\theta = \theta^{-1}$.

The obvious example of an involution is in projective geometry. If λ and λ' are connected by the *symmetrical* relationship $a\lambda\lambda'+b(\lambda+\lambda')+c = 0$ then the mapping $\lambda \to \lambda'$ is an involution: here θ is the mapping of λ into λ' and θ^{-1} maps λ' into λ, and these are the same.

Another example is given by inversion: if P and P' are inverse points then the mapping of P into P' is the same as that of P' into P. If we invert twice (map by θ^2) we get back to the original figure.

Worked exercises

1. Prove that the set of reals $0<x<1$ may be put into 1-1 correspondence with the set $0<x<2$ (i.e. these sets have the same cardinal number).

If x is in the set $0<x<1$ define a mapping by $x\theta = 2x$. This is a mapping of the set $0<x<1$ into the set $0<x<2$ and is clearly both onto and 1-1, and so gives a 1-1 correspondence between the sets.

Exercises 9A

1. Is reciprocation in a plane an involution when considered as operating on: (i) the set of points only, or (ii) the set of points and lines?

Are the functions in **2–8** mappings? If not state why not.

2. $x \to \sqrt[3]{x}$ from the reals into the reals.

3. $x \to \sqrt[3]{x}$ from the reals into the complex numbers.

4. $x \to +\sqrt{x}$ from the reals into the complex numbers.

5. $x \to +\sqrt{x}$ from the reals into the reals.

6. $x \to +\sqrt{x}$ from the positive reals into the reals.

7. $x \to \tan x$ from the reals into the reals.

8. $x \to \cos x$ from the reals into the reals.

Are the mappings in **9–19** onto, 1-1 or neither or both? If they are both onto and 1-1 give the inverse mapping. If they are not onto give the image of the mapping (i.e. $A\theta$ if $\theta: A \to B$).

9. $x \to x^3$ of the reals into the reals.

10. $x \to x^4$ of the reals into the positive or zero reals.

11. $x \to x^3$ of the positive reals into the reals.

12. $x \to x^2$ of the complex numbers into the complex numbers.

13. $x \to \sin x$ of the reals into the reals.

14. $x \to \tan x$ of the reals of modulus $< \frac{1}{2}\pi$ into the reals.

15. $x \to \sin x$ of the reals of modulus $< \frac{1}{2}\pi$ into the reals.

16. A human $\to$ its father as a mapping of the set of all humans into itself.

17. A parent $\to$ its eldest child as a mapping of the set of all parents into the set of all humans.

18. $P \to d(P)$ of polynomials into the positive or zero integers.

19. The permutation of three numbers which sends $1, 2, 3 \to 2, 3, 1$, respectively.

20. Give a 1-1 correspondence between the positive reals and all the reals.

21. Considering the mapping of the days of the year into the days of the week on which they fall, when will the mapping corresponding to the year 1901 first be repeated? When will that corresponding to 1904 first be repeated?

22. If $\theta: A \to B$ and $\phi: B \to A$ are such that $\theta\phi = I_A$, prove that θ is 1-1 and ϕ is onto.

23. In no. **22** if θ is given onto show that ϕ is 1-1 and $\phi = \theta^{-1}$.

24. If $\theta: A \to B$ and $\phi: B \to A$ are such that $\theta\phi = I_A$, $\phi\theta = I_B$, show that θ and ϕ are both 1-1 and onto, and that $\phi = \theta^{-1}$.

Exercises 9B

1. Prove that a countable set is the 'smallest' infinite set in the sense that any infinite set contains a countable subset.

2. Prove that the set of all two-dimensional vectors with integer coefficients is countable.

3. If A is countable and B is finite prove that $A \cup B$ is countable.

4. If A and B are both countable show that $A \cup B$ is countable.

5. Show that the relation of being able to be put in 1-1 correspondence is an equivalence relation between sets, and deduce that all sets fall into equivalence classes, all the sets in each class having the same cardinal number.

6. Prove that no finite set can have the same cardinal number as a proper subset of itself, but give an example to show that this is not necessarily true for infinite sets.

10
THE FUNDAMENTAL LAWS OF ALGEBRA

10.1. The algebra of numbers: the four rules

In elementary algebra we operate with numbers, combining them in various ways according to certain rules. For example, we add two numbers to obtain another, called their sum, or we divide one number by another to form a quotient, and these processes obey such laws as, for example, that $a+b = b+a$. In arithmetic we are interested in the properties of individual numbers, e.g. that $2+3 = 4+1$, but in algebra we replace these specific results by general ones, using letters to stand for any number. Thus $(x+y)(ax+by) = ax^2+(a+b)xy+by^2$ whatever the values of x, y, a and b. This result, furthermore, depends only on the operations of addition and multiplication and the laws which these operations satisfy: the fact that our letters stand for numbers does not enter into the demonstration at all, although it is understood in elementary work.

Elementary algebra depends on surprisingly few operations. When we are dealing with real numbers the basis of all our work is the four rules of addition, subtraction, multiplication and division, together with the idea of order, that is of 'greater than' or 'less than'. When complex numbers are introduced we have to dispense with the idea of order and base everything on the four rules. Thus x^2 merely means $x \times x$, and $\sqrt{x}$ is a number which when multiplied by itself gives x.

Note that we are dealing with algebra, which uses finite processes only, and not with analysis, which involves the infinite. Limits, infinite series, functions such as the trigonometric functions which are defined by means of series, and the whole field of the calculus belong to analysis, and these certainly need other concepts and operations.

Thus apart from the idea of order, the four rules or operations are the fundamental tools of our algebraic work with numbers. They are used according to certain laws—the so-called 'Fundamental Laws of Algebra'—and in the present chapter we will investigate these laws and some of their elementary consequences. We will not deal here with order, since this is a different kind of concept to the rules, being more a property of real numbers than a method of operating with them.

If we look at some of the sets of numbers mentioned in previous chapters we see that all four of the rules do not always apply. Thus the positive integers can be added and multiplied but not always subtracted or divided; when we introduce the negative integers subtraction is always possible, while the set of rationals has all four operations. Again, some or all of the rules apply to sets other than number sets. Thus residue classes always have the first three, and have all four when they are modulo a prime number. Vectors may be added or subtracted, but there is no satisfactory definition of multiplication in general.

We will later be operating with some or all of the four rules, according to some or all of the fundamental laws of algebra, in sets other than those of numbers. Hence when we study the processes in the next few sections, we must bear in mind that the results are equally valid in these more general sets.

A set which has all four of the rules, used according to the usual laws, is called a *field*, and all the work of this chapter will apply to any field. Obvious fields are the sets of the rationals reals and complex numbers. More interesting in many ways are the finite fields formed by the sets of residues modulo p, where p is prime. We recall from §6.3 that all sets of residues admit of addition, subtraction and multiplication, but that division is only unique when the modulus is prime. The arithmetic modulo p forms a field of p elements.

In the rest of the chapter, when we use the word 'number' we mean an element of any field (in certain cases the work will of course apply to other sets as well, but we will restrict our thinking for the moment so that we may concentrate on the four rules themselves). The application of the results to the fields of the rationals, reals and complex numbers is quite obvious from elementary algebra, but that to the finite fields of residues is often more interesting, and we will sometimes use these to illustrate the work. The reader should revise the relevant parts of chapter 6.

10.2. The laws of addition

The simplest process of elementary algebra is that of addition. If we are given any two numbers x and y we form a third by adding them together and call this number their 'sum', denoting it by $x+y$. In effect we have a third, unique, number *associated with* any pair of given ones, and this process of forming the sum obeys the two laws A1 and A2 below. Until we introduce the ideas of subtraction and multiplication these two laws are all that we use in investigating the addition process in algebra: the idea of addition as the combining of two heaps of matches or two pieces of a cake that we meet in our early work in arithmetic is not relevant to the *algebraic* work which, as we have said before, deals with arbitrary numbers.

The two laws of addition are:

A1. The Commutative Law of addition. $x+y = y+x$ *for all pairs of numbers* x *and* y.

A2. The Associative Law of addition. $(x+y)+z = x+(y+z)$ *for any three numbers* x, y *and* z.

The Commutative Law

This states that in forming the sum it does not matter which number of the pair we take first. The symbol ' = ' means, of course, 'is the same number as'. We think of the sum as a number associated with an ordered pair (x, y), and the Commutative Law states that the ordering is not, in fact, significant.

When dealing with sets of numbers the Commutative Law is obviously true, but when we extend our idea of sum to more general sets it turns out that it is not always satisfied, and furthermore that this does not matter very much—we can have perfectly good algebra without it. Anticipating a little, we will find that the product of two numbers behaves much the same as the sum, and we remember (§5.5) that the product of two quaternions is not commutative in general. Neither is the vector product of two three-dimensional vectors.

In the following work it will be noticed that the Commutative Law is seldom used, and in the succeeding chapters it will not be assumed to hold unless this is explicitly stated. To anticipate again, we do not need it until we come to connect addition and multiplication by the Distributive Law, and even then we do not need the Commutative Law of multiplication.

The Associative Law

Addition is initially defined for two numbers only, and to extend it profitably to three or more we need the Associative Law. This means that if we form the sum of x and y *in that order* (remember that we are not assuming A1) and then the sum of our answer (which is itself a number) and z, we get the same result as if we had formed the sum of x and the number which is the sum of y and z.

We now define the sum of x, y and z in that order to be the number given by either side of the equation in A2, and we write it as $x+y+z$. There is no ambiguity in the omission of brackets, because of the Associative Law.

A2 allows us to extend the idea of addition to more than three numbers. It can be readily shown by repeated applications that $x+y+z+w$ has the same value wherever brackets are inserted (e.g. $(x+y)+(z+w) = x+((y+z)+w)$) and so can be written without ambiguity, and the same is true for any finite number of terms. We must of course keep the terms in the same order unless A1 is true. For instance, it would not be true to say that $(x+y)+z = x+(z+y)$ in a system where A1 does not hold.

When all the terms are the same we write $x+x$ as $2x$, $x+x+x$ as $3x$ and so on, and in this case the order does not matter, of course. Note that nx does not mean n multiplied by x, at any rate at present, but is merely a shorthand for the sum of n x's. It is defined only when n is a positive integer.

Unlike the Commutative Law, the Associative Law is vital to most algebraic structures. Fortunately it usually holds in practice, though there are examples where this is not so (see §10.7).

As an example of the Associative Law in the arithmetic modulo 7, consider $4+5+3$ (using ordinary notation and not the bold type of chapter 6—we will also use the ordinary equals sign instead of the identity sign that is sometimes used for congruences)

$$4+5 = 2 \text{ (modulo 7) and } 2+3 = 5,$$
$$5+3 = 1 \text{ and } 4+1 = 5 \text{ as before.}$$

10.3. Subtraction and the zero

The process of subtraction is obviously closely connected with that of addition, and arises from our wish to solve such equations as $x+a = b$. In order to find x we subtract a from both sides, and this means *adding* $-a$ to both sides. Thus we can define subtraction in terms of addition provided we have a definition of $-x$. But $-x$ is the number which when added to x gives zero, and we are led to a definition of zero *in algebraic terms* as a first step towards our definition of subtraction.

0 is the number which, when added to any number x, leaves it unchanged. It is this property which gives it its algebraic importance, and we postulate the existence of such a number in the third law of addition:

A3. The existence of a zero. *There is a number, called the zero and denoted by* 0, *which has the property that*

$$x+0 = x \quad \textit{and} \quad 0+x = x \quad \textit{for all numbers } x.$$

Note that unless we assume A1 we must give both the equations above.

The question arises whether or not it is possible for two or more numbers to have the property given. It is obviously impossible when we are dealing with numbers, but we are now investigating from the point of view of the fundamental laws alone, and it is conceivable that two zeros could exist in some system other than numbers. In fact this is not possible.

Theorem 10.3.1. *The uniqueness of zero.*

There cannot be two different numbers both having the zero property.

Suppose that there were two numbers e and f with the property. Then since e has the property $e+x = x$ for all x. Therefore $e+f = f$. Since f is a zero $y+f = y$ for all y, and so $e+f = e$. Thus $e = f$, since the sum of e and f is unique.

Having defined 0 we are now in a position to postulate the existence of negatives.

A4. The existence of negatives. *Corresponding to each number x there is a number called the negative of x and written $-x$ which satisfies*

$$x+-x = 0 \quad \textit{and} \quad -x+x = 0.$$

For the moment the '$-$' is merely a symbol—it has nothing to do with subtraction, which will be defined in terms of our negatives. The law states that a negative exists for *each* number, negative ones as well as positive ones. We are not here concerned with constructing negative numbers from the positives, rather we are given some set of numbers, say the reals or the complex numbers, and we are stating that every number of the set has its negative (satisfying the equations in A4) within the set: this existence of a negative to every member is necessary for subtraction to be always possible, and is in fact also sufficient for this.

In the arithmetic modulo p, the negative of a number x is $p-x$, for the sum of these is p (which $= 0$ modulo p). Thus modulo 5 we have $-1 = 5-1 = 4$, $-2 = 3$, $-3 = 2$ and $-4 = 1$, while $-0 = 5-0 = 5 = 0$. Note also that numbers and their negatives occur in pairs, thus $2 = -3$ and $3 = -2$: in other words $-(-x) = x$. This, and the fact that $-0 = 0$, are true generally, as shown in theorems 10.3.3 and 10.3.4 below.

We would like the negative to be unique, so that the symbol $-x$ has a definite meaning, and this in fact is so.

Theorem 10.3.2. *The uniqueness of the negative.*

For any x there cannot be two numbers with the property possessed by $-x$ in A4.

Suppose that both y and z have the property. Then

$$\begin{aligned}(y+x)+z &= 0+z \quad \text{by A4 since } y \text{ is a negative of } x\\ &= z \quad \text{by A3.}\end{aligned}$$

And

$$y+(x+z) = y+0 \quad \text{by A4 since } z \text{ is a negative of } x$$
$$= y \quad \text{by A3.}$$

But by A2 $(y+x)+z = y+(x+z)$. Therefore $y = z$.

We now write $x-y$ to mean $x+-y$ and we have defined subtraction, possessing all the usual properties, in terms of addition plus the laws A3 and A4.

Theorem 10.3.3. *The negative of* 0 *is* 0, *i.e.* $-0 = 0$.

$$0+0 = 0 \quad \textit{by } \text{A3.}$$

Thus 0 satisfies the equations $0+x = 0$ and $x+0 = 0$, which are the equations satisfied by the negative x of 0. But the negative is unique, and hence 0 *is* the negative of 0.

Theorem 10.3.4. *The negative of* $-x$ *is* x, *i.e.* $-(-x) = x$.

The negative y of $-x$ satisfies the equations

$$-x+y = 0 \quad \text{and} \quad y+-x = 0.$$

But by A4 x itself satisfies these equations, and so must be the negative, since the latter is unique.

Theorem 10.3.5. *The Cancellation Law of addition.*

$$x+y = x+z \Rightarrow y = z,$$
$$y+x = z+x \Rightarrow y = z.$$

If $x+y$ and $x+z$ are the same number, then if we form the sum of $-x$ and each of them we must arrive at the same answer, i.e.

$$-x+(x+y) = -x+(x+z),$$

i.e.

$$(-x+x)+y = (-x+x)+z \quad \text{by A2,}$$
$$0+y = 0+z \quad \text{by A4.}$$

Thus

$$y = z \quad \text{by A3.}$$

Similarly, for the second part, by adding $-x$ to the right on both sides.

(Note that we cannot, without A1, deduce anything from $x+y = z+x$.)

The above theorem shows us that to write out a proof fully, putting in all steps, takes a fair amount of space. We must be careful not to assume results which seem obvious by reason of their familiarity in our previous algebraic work, but which need justification when we are working from the laws alone. When we become familiar with the new habits of thought we can, however, take the more obvious steps for granted, and we soon learn to shorten the work considerably. This must not be done prematurely, and it is essential at this stage to insert each step, stating which law we are using.

Theorem 10.3.6. *If* $x+a = b$ *then* $x = b-a$. *If* $a+x = b$ *then* $x = -a+b$.

If $x+a = b$ we have

$$(x+a)-a = b-a \quad \text{(i.e. } (x+a)+-a = b+-a).$$

Hence

$$x+(a-a) = b-a \quad \text{by A2,}$$

$$x+0 = b-a \quad \text{by A4,}$$

$$x = b-a \quad \text{by A3.}$$

Similarly for the second part.

Unless we assume A1 we cannot deduce that $x = b-a$ from the equation $a+x = b$: we must be careful to add either on the right or the left, doing the same on both sides of our equation. With this limitation we see that this theorem justifies the elementary algebraic process of 'taking a term to the other side of an equation and changing its sign' which is better described as 'adding (or subtracting) the same term to both sides of an equation'. Of course it follows that $x-a = b \Rightarrow x = b+a$, by use of theorem 10.3.4.

Since the converse of theorem 10.3.6 is also true, we see that there is a unique solution to each of the equations

$$x+a = b \quad \text{and} \quad a+x = b.$$

10.4. The laws of multiplication

In § 10.3 we have shown that subtraction may be defined simply in terms of addition. When we come to the third of the four rules, that of multiplication, we are introducing an essentially new process. It is true that multiplication of x by an integer can be interpreted as the adding together of that many x's, and multiplication by a rational number can be defined by an extension of this idea and its inverse, but the product of two irrational or complex numbers cannot be explained so simply, and product may be defined in much more general sets than the number sets. It is necessary therefore to treat multiplication as a second process of operating with numbers, distinct in the first instance from addition, but linked to the latter by the Distributive Law of §10.5.

It is important to notice in this section and in §10.6 that, apart from one or two minor differences to be discussed later, the laws of addition and multiplication are identical apart from notation.

Thus corresponding to any two numbers x and y, taken in that order, there is a unique third number associated with the pair, called the 'product' of x and y and written xy. This product satisfies the following two laws:

M1. The Commutative Law of multiplication. *$xy = yx$ for all pairs of numbers x and y.*

M2. The Associative Law of Multiplication. *$(xy)z = x(yz)$ for any three numbers x, y and z.*

The Commutative Law

It does not matter in which order we take the pair of numbers in forming their product.

In the same way as the Commutative Law of addition was not used in proving the theorems of §10.3, so this law M1 will not be used in the subsequent work and will not be assumed to hold unless this is specifically stated. It is not necessary for our algebra, even when we connect our processes of addition and multiplication by the Distributive Law (when the law A1 becomes indispensable), and it does not hold for many of the processes which we will wish to think of as products. Remember that it is the first law to be discarded when we extend our ordinary complex numbers to quaternions.

The Associative Law

This defines without ambiguity the product of three, or more, numbers, and enables us to write both sides of the equation given in its enunciation as xyz. Similarly, for $xyzt$, etc., where it does not matter in which order we perform the products of pairs, provided that we always keep the individual numbers in the same order. Thus xyz is not the same as yzx or yxz. As for addition, the law M2 is vital for any algebra involving product.

We write xx as x^2, xxx as x^3 and so on, and we have the Index Laws that

$$x^m . x^n = x^{m+n}, \quad (x^m)^n = x^{mn}.$$

In the arithmetic modulo 7 consider the product 5.6.4

$$5.6 = 30 = 2 \quad \text{and} \quad 2.4 = 8 = 1.$$

Also $6.4 = 24 = 3$ and $5.3 = 15 = 1$ as before, demonstrating the Associative Law in this case.

10.5. The Distributive Law

So far our multiplication has been treated as a completely separate process to addition. This is often necessary—often we have a product defined but not a sum. We sometimes wish, however, to introduce both ideas, as is the case when dealing with numbers, and we then need to connect them in some way. The law of connection is the Distributive Law:

D1. The Distributive Law.

$$(x+y)z = xz+yz,$$

$$x(y+z) = xy+xz$$

for any three numbers x, y and z.

Notice that we need both parts of the law since we are not assuming M1. Note also that the Distributive Law makes a definite distinction between sum and product, for if we interchanged these the law would read

$$xy+z = (x+z)(y+z),$$

$$x+yz = (x+y)(x+z)$$

and these of course are not true for numbers and it would not be fruitful to assume them for other sets.

We can immediately extend the Distributive Law as follows:

Theorem 10.5.1. $(x+y+z)w = xw+yw+zw,$

$$x(y+z+w) = xy+xz+xw.$$

For $(x+y+z)w$ means $((x+y)+z)w$ and this equals

$$(x+y)w+zw \quad \text{by D1}.$$

Thus

$$\begin{aligned}(x+y+z)w &= (x+y)w+zw\\ &= (xw+yw)+zw \quad \text{by D1}\\ &= xw+yw+zw \quad \text{by definition of the R.H.S.}\end{aligned}$$

Similarly, for the second part of the theorem.

Note that if we expressed $(x+y+z)w$ in the form $(x+(y+z))w$ the proof would still apply in the same way, showing us that there is no contradiction in this case between the laws A2 and D1.

Similarly we can prove that

$$((x+y)z)w = (x+y)(zw) = xzw+yzw.$$

Now suppose that we wish to extend D1 by considering $(x+y)(z+w)$. We can think of this in two ways:

(*a*) $$\begin{aligned}(x+y)(z+w) &= x(z+w)+y(z+w)\\ &\qquad \text{by the first part of D1}\\ &= xz+xw+yz+yw\\ &\qquad \text{by the second part of D1;}\end{aligned}$$

(*b*) $$\begin{aligned}(x+y)(z+w) &= (x+y)z+(x+y)w\\ &\qquad \text{by the second part of D1}\\ &= xz+yz+xw+yw\\ &\qquad \text{by the first part of D1.}\end{aligned}$$

Thus we have two different expressions for the expansion of the L.H.S., unless we assume the Commutative Law of addition, in which case we can interchange the order of the two middle terms and overcome the difficulty. Hence if the Distributive Law is not to lead to inconsistence we must assume A1, and in future *we will always suppose this to apply whenever we have the*

Distributive Law, i.e. whenever we are working with addition and multiplication at the same time. There is still no need to assume M 1, and we will not do so. It will be seen in both (*a*) and (*b*) above that x and y always remain on the left of the products, and z and w on the right.

Such extensions as the expansion of $(x+y)(z+t)(u+v)$ are now straightforward, bearing in mind that the order in which we write the terms does not now matter, but the order of the letters in each term must not be changed: x or y comes first, followed by z or t with u or v last.

Theorem 10.5.2. $x0 = 0x = 0$ *for any* x.

Let a be any number. Then

$$x(a+0) = xa+x0 \quad \text{by D 1.}$$

But $$a+0 = a \quad \text{by A 3.}$$

Therefore $$x(a+0) = xa$$

and so $$xa+x0 = xa = xa+0 \quad \text{by A 3.}$$

Hence by the Cancellation Law of addition $x0 = 0$.

Similarly $(a+0)x = ax+0x$. And $(a+0)x = ax = ax+0$. Hence $0x = 0$.

Note that in the arithmetic modulo p this states that x multiplied by a multiple of p (which $= 0$ modulo p) leads to a product which is also a multiple of p.

Theorem 10.5.3.

$$(-x)y = -(xy), \quad x(-y) = -(xy). \quad (-x)(-y) = xy,$$

$$\begin{aligned} xy+(-x)y &= (x+-x)y \quad \text{by D 1} \\ &= 0y \quad \text{by A 4} \\ &= 0 \quad \text{by theorem 10.5.2.} \end{aligned}$$

Similarly $(-x)y+xy = 0$, and so $(-x)y$ is the inverse of xy; i.e. $(-x)y = -(xy)$.

$x(-y)+xy = x(-y+y) = x0 = 0$ as above. Similarly

$$xy+x(-y) = 0 \quad \text{and} \quad x(-y)$$

is the inverse of xy.

By the first part

$$(-x)(-y) = -(x(-y))$$
$$= -(-(xy)) \quad \text{by the second part}$$
$$= xy \quad \text{by theorem 10.3.4.}$$

10.6. Division and the unity

In a way similar to that which we used to define subtraction in terms of addition we will now derive the process of division from that of multiplication. The first step, as before, is to postulate the existence of a neutral number which does not alter other numbers when multiplying them. This is clearly the number 1, called the unity.

M3. The existence of a unity. *There is a number, called the unity and denoted by* 1, *which has the property that*

$$x1 = 1x = x \quad \textit{for all numbers } x.$$

We prove the uniqueness of the unity by a theorem analogous to theorem 10.3.1.

Theorem 10.6.1. *The uniqueness of the unity.*

There cannot be two numbers both having the unity property.

Suppose that there were two numbers e and f with the property. Then since e has the property $ex = x$ for all x. Therefore $ef = f$. Since f is a unity $yf = y$ for all y, and so $ef = e$. Thus $e = f$, since the product of e and f is unique.

We now lay down the existence of reciprocals, or, as we will call them, 'inverses'.

M4. The existence of inverses. *Corresponding to each number* x, except 0, *there is a number called the inverse of* x *and written* x^{-1}, *which satisfies*

$$xx^{-1} = 1 \quad \text{and} \quad x^{-1}x = 1.$$

We are compelled to make an exception of 0, since by theorem 10.5.2 the product of 0 with any number is 0 and so cannot be 1. (0 and 1 are different, since if not we would have $1x = x$ and $1x = 0x = 0$ so that all our numbers x would be 0 and our whole system would be trivial, consisting of this one number only.)

We recall that in §6.3 we defined s/r, where r and s are residues, to be a solution of $rx = s$. If $s = 1$ this gives a definition of $1/r$, or, as we prefer to write, r^{-1}, in line with M4. We showed that division was always possible for a prime modulus, and proved it unique by a method using the finite nature of the field (if $rx = s$ has a solution for x for all s then they must be different and so each s can occur from one value of x only). To prove uniqueness in general we cannot assume finiteness and we must use a method analogous to that for negatives.

Theorem 10.6.2. *The uniqueness of the inverse.*

For any x there cannot be two numbers with the property possessed by x^{-1} in M4.

Suppose that both y and z had the property. Then

$$(yx)z = 1z \quad \text{by M4 since } y \text{ is an inverse of } x$$
$$= z \quad \text{by M3.}$$

And
$$y(xz) = y1 \quad \text{by M4 since } z \text{ is an inverse of } x$$
$$= y \quad \text{by M3.}$$

Therefore by use of M2, $y = z$.

We always denote the inverse by x^{-1} and not $1/x$ for the following reason. We wish to differentiate between yx^{-1} and $x^{-1}y$, and with the notation $1/x$ we would have to write both these as y/x, unless we were to use $y(1/x)$ and $(1/x)y$, which would make for clumsiness. In fact the notation x^{-1} helps us to remember that the inverse is an ordinary number, and we soon become familiar with its use.

Division by a number means multiplication by its inverse.

In the arithmetic modulo 5,

$$1^{-1} = 1,\ 2^{-1} = 3 \text{ (since } 2.3 = 6 = 1),\ 3^{-1} = 2 \text{ and } 4^{-1} = 4.$$

Note that 1 is its own inverse, as is 4 in this case, while 2 and 3 are the inverses of each other. The next two theorems show that these properties hold generally. (The fact that 4 is its own inverse modulo 5 shows that there may be other numbers besides 1 with this property, but 1 is the only number that *always* possesses it—modulo 7 the inverse of 4 is 2.) Modulo 5, division by 2 means multiplication by 3, for example,

$$\tfrac{3}{2} = 3.3 = 4,$$

where of course $\frac{3}{2}$ means the solution of $2x = 3$, and we see that 4 is indeed a solution of this, and is of course unique. The uniqueness of division is shown in general by the Cancellation Law, proved generally in theorem 10.6.5, and the unique quotient of b by a is given in theorem 10.6.6. In these two latter theorems we do not assume the Commutative Law of multiplication, and thus need to consider multiplication on each side, giving two cases whose proofs are similar.

Theorem 10.6.3. *The inverse of* 1 *is* 1, *i.e.* $1^{-1} = 1$.

$1.1 = 1$ by M3. Thus 1 satisfies the equations $1x = 1$ and $x1 = 1$, and so is the inverse of 1, since this inverse is unique.

Theorem 10.6.4. *The inverse of* x^{-1} *is* x, *i.e.* $(x^{-1})^{-1} = x$.

The inverse y of x^{-1} satisfies the equations

$$x^{-1}y = 1 \quad \text{and} \quad yx^{-1} = 1.$$

But by M4 x itself satisfies these equations, and so must be the inverse, the latter being unique.

Theorem 10.6.5. *The Cancellation Law of multiplication.*

$$\left.\begin{aligned} xy = xz \Rightarrow y = z \\ yx = zx \Rightarrow y = z \end{aligned}\right\} \quad \textit{unless } x = 0.$$

If $xy = xz$ then, provided $x \neq 0$, so that x^{-1} *does* exist, $x^{-1}(xy) = x^{-1}(xz)$.

Or, by M2, $(x^{-1}x)y = (x^{-1}x)z$.

Therefore $1y = 1z$ by M4.

And so $y = z$ by M3,

and similarly for the second part.

Note. If $x = 0$ it has no inverse and the above proof breaks down. Of course in this case $xy = xz = 0$ whatever the values of y and z. Hence *we cannot divide by zero.*

Theorem 10.6.6. *The equation $ax = b$ has the unique solution $x = a^{-1}b$, unless $a = 0$, in which case there is no solution unless $b = 0$, when every x is a solution.*

(*a*) Suppose $ax = b$ with $a \neq 0$. Then $a^{-1}ax = a^{-1}b$. But $a^{-1}ax = 1x = x$ and so we must have $x = a^{-1}b$. Hence if there is a solution it must be $x = a^{-1}b$.

(*b*) Suppose $x = a^{-1}b$. Then $ax = aa^{-1}b = 1b = b$. Hence $x = a^{-1}b$ is a solution, and by (*a*) it is the unique solution.

If $a = 0$ then $ax = 0$ for all x and the second part follows.

Similarly, we may prove that the equation $xa = b$ has the unique solution $x = ba^{-1}$, unless $a = 0$.

10.7. The laws for sets other than numbers

Although in the previous sections we have always understood our elements x, y, etc., to be numbers of one kind or another, we have hinted that this is not necessary, and we have not used this fact in our work, except to notice that numbers do in fact obey the fundamental laws.

All our work so far in this chapter depends on the fundamental laws being assumed to hold, and so it can be done with any elements which obey them. Let us investigate exactly what we mean by this.

Suppose we have a set S within which we have some process which satisfies A1 and A2. This means that given any two elements of our set, in a definite order, there is a third element which is unique and must be in the set S, associated with them, which element we write as $x+y$ if x and y are our original elements. Then this process must satisfy A1 and A2 whichever elements we take in the set. The fact that the process satisfies A1 means that the element associated with the pair (x, y) is the same as that associated with (y, x). ('$=$' means 'is the same element as'). Similarly for A2, which must hold whatever three elements we start with provided these are in S. If we have such a process we call it addition and use the sum notation, not because this means exactly the same as with numbers (in the set of residues modulo 4 we have $3+2 = 1$, *not* 5), but because algebraically it behaves in the same way and since we are familiar with this notation it makes our work easier, once we realise exactly what we may assume.

Unless our process obeys the laws A3 and A4 also we cannot produce much algebra. If we have these two laws, even if A1 does not hold, we can build a surprisingly large structure on to our set. The law A3 means that there is an element 0, in the set S, which satisfies A3 in connection with any x in the set. That is, the element associated with either the pair $(x, 0)$ or the pair $(0, x)$ must be the element x. If we assume *only* the two laws A2 and A3 we see as in theorem 10.3.1 that we can prove that there cannot be more than one element with the given property. The law A4 says that there exists, within our set, an element which gives the stated results when associated with x: it will be a different element for each x. If we assume A1, A3 and A4 we can prove the remaining theorems in §10.3.

We now introduce a second process called multiplication, though we need not already have an addition defined unless we wish to apply D1. The multiplication must satisfy M2 for it to be useful, and *may* satisfy M1. Again, it says that to any two elements in S, in a particular order, there is an associated element, which must be in S and which is unique, called the product and satisfying the relevant rules.

If we have both a sum and product then D1 must be satisfied in order for them to combine fruitfully, and in this case we found that we must have A1, though not necessarily M1. We can then prove the theorems in §10.5.

If we have a product, irrespective of whether there is a sum defined, we must have M3 and M4 true if we wish to 'divide'. If there is no sum, so that there need be no zero, then we have seen that inverses can always exist and are unique, while if we have a sum and product together, as happens in ordinary algebra, we have to make an exception of 0.

We will later meet many examples of sets which contain processes which obey some or all of the fundamental laws, but we have already met a few in previous chapters. If we have the set of two-dimensional vectors, defined as displacements, then the 'sum' is that displacement obtained by performing first one, then the second, displacement. We saw that it satisfied A1–4, but this needed proving and was not obvious. Multiplication was not easy to define. For the complex number multiplication of two-dimensional vectors we clearly have all the fundamental laws. The scalar product of three-dimensional vectors is not even a vector, and is not in our set, being therefore useless for our present algebraic purposes, though useful of course in other contexts. The vector product is a vector, but is not commutative, since $\mathbf{x} \wedge \mathbf{y} = -(\mathbf{y} \wedge \mathbf{x})$. This is not important, but also it is not associative, which prevents it being a useful *algebraic* concept. The fact that it is not associative can be seen from the following special case. For most sets of 3 vectors we would find that M2 did not hold, but the given counter example is sufficient to disprove the law.

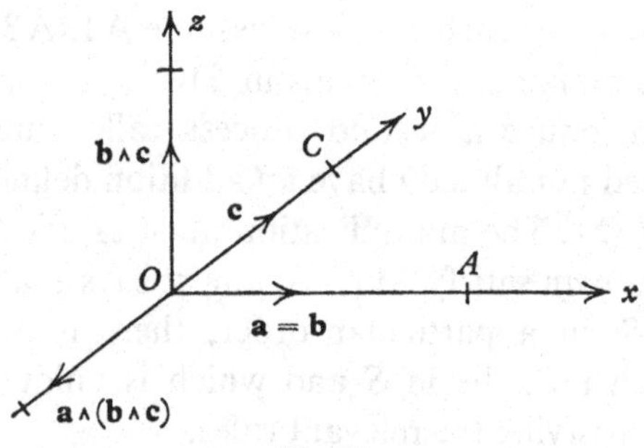

Fig. 39

Theorem 10.7.1. *The vector product is not associative.*

Let
$$\mathbf{a} = \mathbf{b} = (1, 0, 0), \quad \mathbf{c} = (0, 1, 0).$$
Then since $\mathbf{a} = \mathbf{b}$, $\mathbf{a} \wedge \mathbf{b} = \mathbf{0}$.

Hence $(\mathbf{a} \wedge \mathbf{b}) \wedge \mathbf{c} = \mathbf{0} \wedge \mathbf{c} = \mathbf{0}$.

But $\mathbf{b} \wedge \mathbf{c} = (0, 0, 1)$ and so
$$\mathbf{a} \wedge (\mathbf{b} \wedge \mathbf{c}) = (1, 0, 0) \wedge (0, 0, 1)$$
$$= (0, -1, 0), \quad \text{which is not } \mathbf{0}.$$

Another example of non-associativity is found when dealing with exponents. Consider
$$(3^3)^3 = 27^3 = 3^9,$$
$$3^{(3^3)} = 3^{27}.$$
Thus $(3^3)^3$ does not equal $3^{(3^3)}$.

10.8. The similarity between sum and product

If we look at the laws A 1–4 and M 1–4, and at the conclusions drawn from them, we see that apart from notation they are identical except for two things. We will look first at the similarity and then discuss the differences.

We remember that when defining either the sum or the product we were interested in the fact that there was an element associated with any ordered pair of elements—in one case we called this the sum and wrote $x+y$, while in the other we called it the product and wrote xy. With this difference in notation the Commutative and Associative Laws were identical. The other pairs of laws also become the same if we replace 0 in A 3 by 1 in M 3, and further if $-x$ in A 4 becomes x^{-1} in M 4. The theorems deduced from each set of laws are also identical (except for the exceptional character of 0 when introducing inverses). Thus apart from notation both the results and the proofs of theorems 10.3.1–10.3.6 are the same as those of theorems 10.6.1–10.6.6, respectively. Thus as far as the above work is concerned it is immaterial which notation we use, and this is often so in work involving these processes.

The two differences are:

(*a*) In the Distributive Law;

(*b*) In the fact that 0 has no inverse, and the consequences of this.

The first of course is an important difference. It makes a distinction between the two processes of addition and multiplication but of course it applies only when both are present. The second difference, the lack of inverse of 0, arises from theorem 10.5.2, which depends on the Distributive Law. Thus it also applies only when both sum and product are present together in a set, indeed there is no meaning to be attached to 'zero' unless there is a sum, and inverses arise only with products.

Hence we have the conclusion that, provided only one process is present in a set, which satisfies the required laws, it may be given the name and notation either of 'sum' or of 'product'—in the first case we introduce 'zero' and 'negative', while in the second these become 'unity' and 'inverse'. (In fact it is sometimes more convenient to call the unity the 'neutral element' and to designate it by 'e', as will be done in the next chapter.) Apart from the different notation the work and results are the same.

Either sum or product may be, and is, used, but it is more usual to use the product notation than the sum one, since it is simpler to write xy than $x+y$. The sum notation is usually reserved for the special case when the Commutative Law is true —most of the processes that we ordinarily call sum are commutative, and the use of this notation enables us to bear A1 in mind as we work our algebra.

As soon as we introduce two processes we need to have the Distributive Law to connect them, and this makes a definite distinction between sum and product in this case. We now have to modify our definition of inverse to exclude 0, and we must also remember that sum must be commutative to give a consistent algebra, as we showed in §10.5.

10.9. Algebraic structures

In order to use the methods of algebra in a set we must have some process, such as addition or multiplication, connecting our elements of the set. A set which has one or more operations such as these is called an *algebraic structure*.

One of the processes may satisfy A2–4, in which case it may be called addition, though it would often be put in the product notation and be called multiplication. It may satisfy A2 alone, or perhaps A1 and A2, but in this case there is very little we can do with it, unless other operations are also present. We may have two processes, one satisfying A1–4, the other M2–4, or perhaps merely M2, with D1 holding in addition, while M1 may or may not apply.

On the other hand our processes need not be of this type. If we consider the set whose elements are all the subsets of a given set (e.g. an element is any collection of positive integers, this being a subset of the set of all positive integers) then we can define two operations, one as the intersection ($\cap$) of two subsets, the second as the union ($\cup$) of two subsets. As we saw in chapter 3, such processes satisfy laws similar to our fundamental laws, but not quite the same. Again, we could form a new positive integer from two given positive integers x and y by raising x to the power y: we have seen that this is not associative, nor is it commutative, and it is not likely to be a fruitful idea as an algebraic operation.

It is possible that *three* (or more) elements need to be taken in order to define a new element. An example of this would be the formation of $\mathbf{a} \wedge (\mathbf{b} \wedge \mathbf{c})$ for the three-dimensional vectors **a**, **b** and **c**. Another example using four numbers is the cross-ratio $(\lambda_1 - \lambda_3)(\lambda_2 - \lambda_4)/(\lambda_2 - \lambda_3)(\lambda_1 - \lambda_4)$ of the four numbers $(\lambda_1, \lambda_2, \lambda_3, \lambda_4)$. We must expect the laws for these more difficult processes to be correspondingly complicated: for instance, a complete Commutative Law would need 24 cross-ratios to be equal, and does not of course hold in this example, although a partial one, giving six connected values, does.

As is to be expected by far the most fruitful ideas are those involving two elements only, and among these it is found that those which satisfy some of the laws of this chapter are most useful in practice. This is natural, since these laws are precisely those which hold when we are dealing with ordinary numbers, which always remains the most fruitful set to work in. We dealt with the processes of $\cap$ and $\cup$ in chapter 3, and when dealing with vector spaces in volume 2 will introduce a rather different type of process, but on the whole we will be concerned with forming sums and products.

10.10. Some types of structures

A set with a process which satisfies M 2, M 3 and M 4 (without any exceptions, since we do not need to mention 0) is known as a *group*. It may or may not be commutative. The process may be written in the addition notation and then we must have A 2, A 3 and A 4 satisfied, with A 1 possibly true, and indeed usually true when this notation is used.

If we have a set with two processes, one satisfying A 1–4 and the other M 2, with D1 also applying, then we call it a *ring*. The laws M 1 and M 3 may or may not be true. If in addition we have M 1, M 3 and M 4 we call our structure a *field*. (If M 1 is not true it is called a *skew field*.)

Thus roughly speaking a group possesses multiplication and division (or addition and subtraction if we prefer), a ring has addition, subtraction and multiplication, while a field contains all four rules.

Worked exercises

1. Assuming A1 and A2, prove that $(x+y)+(y+x) = 2x+2y$.

$$\begin{aligned}(x+y)+(y+x) &= ((x+y)+y)+x \quad \text{by A2}\\ &= (x+(y+y))+x \quad \text{by A2}\\ &= (x+2y)+x \quad \text{by A2}\\ &= x+(x+2y) \quad \text{by A1}\\ &= (x+x)+2y \quad \text{by A2}\\ &= 2x+2y.\end{aligned}$$

2. Assuming A2, A3 and A4 prove that $-(x+y+z) = -z-y-x$.

$$\begin{aligned}(x+y+z)+(-z-y-x) &= ((x+y)+z)+(-z+(-y-x))\\ &= (((x+y)+z)-z)+(-y-x) \quad \text{by A2}\\ &= ((x+y)+(z-z))+(-y-x) \quad \text{by A2}\\ &= ((x+y)+0)+(-y-x) \quad \text{by A4}\\ &= (x+y)+(-y-x) \quad \text{by A3}\\ &= ((x+y)-y)-x \quad \text{by A2}\\ &= (x+(y-y))-x \quad \text{by A2}\\ &= (x+0)-x \quad \text{by A4}\\ &= x-x \quad \text{by A3}\\ &= 0 \quad \text{by A4.}\end{aligned}$$

Similarly $(-z-y-x)+(x+y+z) = 0$, and so $(-z-y-x)$ is the negative of $(x+y+z)$.

3. Assuming M2, M3 and M4, prove that $axb = ayb \Rightarrow x = y$.

Since $axb = ayb$ we have

$$a^{-1}axbb^{-1} = a^{-1}aybb^{-1}.$$

Then $1x1 = 1y1$ by M2 and M4.

Therefore $x = y$ by M2 and M3.

4. Assuming all laws prove that $(a-b)^2 = a^2-2ab+b^2$.

$$\begin{aligned}(a-b)^2 = (a-b)(a-b) &= a(a-b)-b(a-b) \quad \text{by D1}\\ &= a^2-ab-ba+b^2 \quad \text{by D1, A2 and theorem 10.5.3}\\ &= a^2-ab-ab+b^2 \quad \text{by M1}\\ &= a^2-2ab+b^2.\end{aligned}$$

5. Solve the equation $(x+a)b = bc$, assuming all laws except M1.

$$(x+a)b = bc,$$

Thus $$x+a = bcb^{-1},$$

multiplying both sides on the right by b^{-1} and applying M4 and M3.
Hence $$x = bcb^{-1}-a,$$

adding $-a$ to both sides and applying A4 and A3.

Exercises 10A

In all these exercises state which laws you are using.

Assuming A2 prove the results **1–3**:

1. $((x+y)+z)+w = x+((y+z)+w)$.

2. $x+((y+z)+(w+t)) = ((x+y)+z)+(w+t)$.

3. $x+((y+z)+(w+t)) = ((x+y)+(z+w))+t$.

Assuming A1 and A2 prove the results **4–10**:

4. $(x+y)+z = x+(z+y)$.

5. $(x+y)+z = (z+y)+x$.

6. $((x+y)+z)+w = (y+z)+(x+w)$.

7. $y+((x+z)+w) = ((w+x)+y)+z$.

8. $(x+y)+x = 2x+y$.

9. $2(y+x) = 2x+2y$.

10. $(x+y)+(y+z) = (x+2y)+z$.

11. Which of the results **4–10** can be proved without assuming A1?

12. Prove that $-(-(-x)) = -x$.

13. Prove that $x-0 = x$.

14. Assuming A2, A3 and A4, what is the solution of:
(i) $x+a+b = c$; (ii) $a+x+b = c$; (iii) $a+b+x = c$?

15. Assuming A2, A3 and A4, prove that:
(i) $-(x+y) = -y-x$; (ii) $-(y+x) = -x-y$.

Assuming A2, A3 and A4, which of the following (**16–20**) may we deduce? If the deduction is possible prove it.

16. $x+y+z = w+y+z \Rightarrow x = w$.

17. $x+y+z = x+w+z \Rightarrow y = w$.

18. $a+b+c = d+b+g \Rightarrow a+c = d+g$.

19. $a+b+c = d+g+c \Rightarrow a+b = d+g.$

20. $a+b+c = c+d+g \Rightarrow a+b = d+g.$

Assuming M2 prove the results **21–22**:

21. $(xy)(zw) = x((yz)w).$

22. $(x(yz))(wt) = (xy)((zw)t).$

Assuming M1 and M2 prove the results **23–26**:

23. $(xy)z = (yz)x.$

24. $(xy)(zy) = (xz)y^2.$

25. $(xy)^2 = x^2y^2.$

26. $(xy)(yz) = (xy^2)z.$

27. Prove that $((x^{-1})^{-1})^{-1} = x^{-1}$ $(x \neq 0).$

28. Assuming M2, M3 and M4, prove that:

(i) $(ab)^{-1} = b^{-1}a^{-1}$; (ii) $(ba)^{-1} = a^{-1}b^{-1}$; (iii) $(abc)^{-1} = c^{-1}b^{-1}a^{-1}$; assuming that none of a, b or c is 0.

Assuming M2, M3 and M4, which of **29–37** can be deduced? Prove the result if it is true.

29. $abx = aby \Rightarrow x = y.$

30. $xcd = ycd \Rightarrow y = x.$

31. $xc = cy \Rightarrow x = y.$

32. $cxd = dyc \Rightarrow y = x.$

33. $abc = dbg \Rightarrow ac = dg.$

34. $ax = 1 \Rightarrow x = a^{-1}, a \neq 0.$

35. $abx = 1 \Rightarrow x = a^{-1}b^{-1}, a \neq 0, b \neq 0.$

36. $xab = c \Rightarrow x = cb^{-1}a^{-1}, a \neq 0, b \neq 0.$

37. $xab = a \Rightarrow x = b^{-1}, b \neq 0.$

38. Assuming all laws except M1 prove that $((x+y)z)w = xzw+yzw.$

Prove the results **39–42**, assuming all the laws except M1:

39. $(a+b)^2 = a^2+ab+ba+b^2.$

40. $a^2(b+c) = a^2c+a^2b.$

41. $(a+b)(a-b) = a^2+ba-ab-b^2.$

42. $b^{-1}a+cd^{-1} = b^{-1}(ad+bc)d^{-1}, b \neq 0, d \neq 0.$

Prove the results **43–46**, assuming all the laws including M1:

43. $(a+b)(a-b) = a^2-b^2.$

44. $a^3+b^3 = (a+b)(a^2-ab+b^2).$

45. $(x+y)^3 = x^3+3x^2y+3xy^2+y^3.$

46. $ab^{-1}+cd^{-1} = (ad+bc)b^{-1}d^{-1}, b \neq 0, d \neq 0.$

Solve the equations **47–50**, assuming all the laws except M1:

47. $x(a+b) = c+d$.

48. $axb = c+d$.

49. $(a+b)(x+c) = d$.

50. $a(x+b)cd = g$.

Exercises 10B

1. Suppose **a**, **b**, **c**, ... specify movements on the surface of the earth. If **a** means 'go 1 mile south', **b** means 'go 2 miles north' and **c** means 'go 3 miles east', and if I start on the equator, which of the following equations are true? Give reasons for your answer.

(i) $\mathbf{a}+\mathbf{b} = \mathbf{b}+\mathbf{a}$;
(ii) $\mathbf{a}+\mathbf{c} = \mathbf{c}+\mathbf{a}$;
(iii) $\mathbf{a}+(\mathbf{b}+\mathbf{c}) = (\mathbf{a}+\mathbf{b})+\mathbf{c}$;
(iv) $\mathbf{c}+(\mathbf{a}+\mathbf{b}) = (\mathbf{c}+\mathbf{b})+\mathbf{a}$;
(v) $(\mathbf{a}+\mathbf{b})+(\mathbf{c}+\mathbf{b}) = \mathbf{a}+(\mathbf{c}+2\mathbf{b})$.

2. If **a** means 'go 1 mile south', **b** means 'go $\frac{1}{2}$ mile west' and **c** means 'stay where you are', and if I start at the North Pole, which of the following are true? If any statements are meaningless say so:

(i) $\mathbf{a}+\mathbf{b} = \mathbf{b}+\mathbf{a}$;
(ii) $\mathbf{a}+\mathbf{c} = \mathbf{c}+\mathbf{a}$;
(iii) $\mathbf{a}+(\mathbf{b}+\mathbf{a}) = ((\mathbf{a}+\mathbf{c})+\mathbf{b})+\mathbf{a}$;
(iv) $\mathbf{a}+(\mathbf{c}+\mathbf{a}) = 2\mathbf{a}+\mathbf{c}$;
(v) $\mathbf{c}+\mathbf{b} = \mathbf{b}$.

3. If we have a set S and a product defined in S satisfying M2 only, we say that an element e of S is a *left unity* if $ex = x$ for all x in S, and an element f is a *right unity* if $xf = x$ for all x in S. Show that if there exist both a left unity and a right unity then these are the same element, and deduce that, provided one of each type exists, there cannot be more than one of each. By considering the product defined by $ab = b$ for all a, b in S, show that it is possible to have a product satisfying M2 such that there are many left unities but no right unity.

4. Suppose S has a product satisfying M2 and M3. Then a *left inverse* of x is an element y such that $yx = 1$ and a *right inverse* is defined similarly. Show that if x has both a left and a right inverse these are unique and are the same.

5. Show that no two distinct elements of a set can have the same inverse.

6. With the usual definition of product which of the following sets have a unity and which have an inverse for every element except, possibly, 0:

(i) The positive integers.
(ii) The negative integers.
(iii) All numbers $a+b\sqrt{2}$, a and b integers, positive, zero or negative?

(iv) All numbers $a+b\sqrt{2}$, a and b rational.
(v) All polynomials with rational coefficients.
(vi) All complex numbers of unit modulus.
(vii) All complex numbers bi, b real.
(viii) All rational functions with integral coefficients.

7. Assuming M1 and M2 prove that $(ab)^n = a^n b^n$ for any positive integer n.

8. Assuming M1 and M2 prove that $a^m b^n = b^n a^m$ for any positive integers m and n.

9. Assuming M2, M3 and M4 prove that $(a_1 a_2 \ldots a_n)^{-1} = a_n^{-1} a_{n-1}^{-1} \ldots a_1^{-1}$.

10. Suppose we have a product defined in a set S and satisfying the following laws:
(i) M3;
(ii) $(ab)c = a(cb)$ for all a, b, c in S.
Show that the product satisfies M1 and hence M2 also.

11

GROUPS

11.1. The idea of a group

In the previous chapter we showed that algebra depends ultimately on the validity of a few fundamental laws. Provided that some or all of these hold in a set we can perform normal algebraic operations on the elements of the set, although in cases where some only of the fundamental laws are true we are naturally restricted to operations which depend on these laws only. The laws are concerned with two fundamental processes, addition and multiplication, which behave identically when one alone is present.

The simplest important algebraic structure has just one operation, which we will usually call multiplication, and which satisfies laws M 2, M 3 and M 4. Such a structure is called a *group*, and we see that it is basically a set within which we can multiply and divide. Notice that the set itself is not the group: the latter is given by the set *plus* the product structure, the way in which each pair of elements combine to give their product being an essential part of the group.

Although the basic laws of a group are so few and simple, we can build a surprisingly large amount of algebra with them. Much of this work is extremely complicated, but we will be able to do a little of it in this and the next two chapters. Because the laws are few many structures are groups and so of course any results proved in the general theory are true in such structures. This explains the great importance and wide application of group theory: we merely have to show that a given structure is a group (usually a simple matter) and then we have the whole range of results ready to apply where necessary. As against this we must realise that we are isolating a few properties of our sets, many of which possess much more than we have used, so that we must not expect group theory alone to tell us all that we may wish to know. For instance, we can find out much about the addition and subtraction of three-dimensional vectors by considering them as a group, but this will tell us nothing about the scalar and vector products.

A group then is a set with a product which obeys the Associative Law, and which contains a neutral element (i.e. a unity) and an inverse to every element. Many examples will be given in the next chapter, but we include one or two here for use later in the present chapter.

Example 1. Consider the set of all two-dimensional vectors, denoted by co-ordinates. These form a group under normal vector addition.

For the sum of any two vectors (a_1, a_2) and (b_1, b_2) is the vector (a_1+b_1, a_2+b_2), which is uniquely defined and in the set.

The sum is associative, obviously, since ordinary numbers are and the vectors are merely pairs of numbers.

There is a neutral element (0, 0) in the set, and the inverse of (a, b) is $(-a, -b)$, which is in the set.

(Note here that the product process is normal *addition*, and this can give rise to confusion in terminology: we can use 'sum' or 'product', depending on whether we are referring to the ordinary vector sum or the group product. There would be a case here for using the addition notation throughout, and this would be done in extensive work on this particular set. The confusion is not really disconcerting; after all 'sum' and 'product' mean the same thing in group theory.)

Example 2. We take the set of residue classes modulo 5, excluding the class containing zero. Our product is the ordinary product of residue classes (i.e. if a and b are elements of the classes A and B the product of A and B is that class containing ab).

The set consists of the classes containing 1, 2, 3 and 4, and we denote the classes by these representatives. Then this set with the ordinary product form a group.

The product of any two residues is clearly a residue, and is always in the set, since it cannot be 0. Thus the product of 2 and 3 is 1, and of 3 and 4 is 2. Also the product is unique, and of course associative.

The neutral element is 1, and inverses exist since division is possible (5 is a prime) and so the inverse of A is $1 \div A$. In fact it is easily seen that the inverses of 1, 2, 3 and 4 are respectively 1, 3, 2 and 4.

Example 3. The set of residues modulo 4 forms a group under addition of residues.

The sum of any two residues is a residue of the set and is unique. It is associative, the neutral element is 0 and inverses exist, the inverse of 0 being 0, and the inverse of A being $4-A$, where A is any class other than 0.

Example 4. *The transformation group of an equilateral triangle.* We give an example now of an important type of group, which will be dealt with more fully in the next chapter. This example is given here since it is a fairly simple non-commutative group.

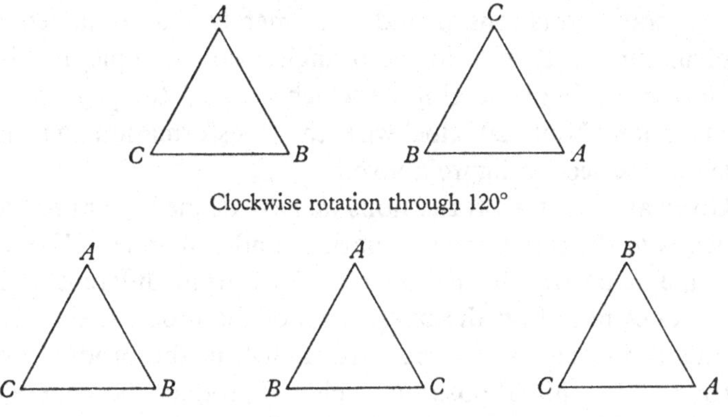

Turning over plus clockwise rotation through 120°

Fig. 40

Suppose we have an equilateral triangle ABC in a certain position. We may move it in several particular ways so that it remains in the same place afterwards, although the positions of the vertices may be changed. Thus we can rotate it clockwise or anticlockwise through any multiple of 120°, or we may turn it over (about an altitude) and then rotate it.

Such a movement which leaves the triangle occupying the same position will be called a transformation; we now consider the set of all transformations of the triangle. There are apparently an infinite number of these, since we may rotate through *any* multiple of 120°. However, the effect of a rotation of 480° is exactly the same as one of 120°, and so we consider these to be the same transformation. Similarly, any two movements which have the same effect are taken to be the same; we are interested in the effect of a movement rather than in the movement itself. As an example the second transformation in the figures above is the same as that consisting of a rotation through 120° anti-clockwise plus turning over about the same altitude (up the paper).

We see that there are in fact just 6 transformations, three obtained by rotation (including leaving the triangle alone), and three by turning over with, possibly, rotation. These 6 form the elements of a set, the set of transformations.

We now define a product of two transformations. The product of the transformations a and b is merely that obtained by applying first a, then b, to the triangle. For example, if a is a turning over about the altitude which runs up the page, and b is a rotation about 120° clockwise, the transformation ab is that given in the second figure above.

Given any two transformations we have defined their product, which is in the set of transformations and is unique. (The fact that the same transformation may arise from different *movements* does not affect the uniqueness of the product, since any of these movements may be used at will in the product and, given a certain initial position, each will produce the same final position.) The product is clearly associative. The neutral element is that transformation which leaves the position unaltered, since the product of this and any transformation a is a. The inverse is the transformation which has the opposite effect, for instance the inverse of a rotation of 120° clockwise is a rotation of 120° anticlockwise.

Thus the set of 6 transformations with product as defined is a group, called the *transformation group* of the triangle. There is no reason for the Commutative Law M 1 to hold, and in fact it does not, as we will see.

We will investigate the structure of this group a little further. The 6 elements are given in figure 41, denoted for the moment by e, the neutral element, and by u, v, w, x, y.

The element u is a rotation of 120° clockwise. Consider uu or u^2. This is a rotation through 120° (clockwise) followed by another 120° rotation, i.e. a rotation through 240° clockwise. We know that it must be an element of the group, and we see that it is in fact v. Now consider u^3. This is a rotation through 360° clockwise, i.e. is e the neutral element. u^4 is of course $u.u^3 = ue = u$, $u^5 = u^2$, and so on.

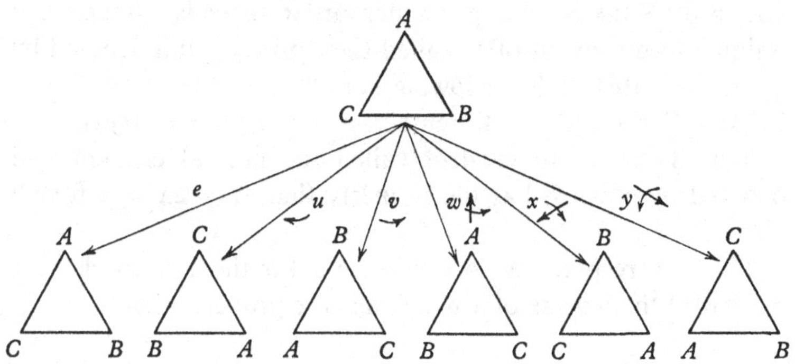

Fig. 41. The curved arrows show the effect of the various rotations.

Now consider wu. This is a turning over about the altitude up the page followed by a rotation of 120° clockwise, i.e. is x. Similarly, wu^2 is y and, of course, wu^3 is $we = w$.

We also have $w^2 = e$. Finally, we think about uw. This is a rotation of 120° clockwise followed by turning over. We easily see that it is the same as y. But wu is x, and is *not* the same as uw. Hence *the group is not commutative*, although of course certain pairs of elements may commute, such as u and v. ($uv = vu = e$.)

We see that we may express all transformations in terms of e and the two elements u and w. Thus our 6 elements are

$$e, \quad u, \quad v = u^2, \quad w, \quad x = wu, \quad y = wu^2.$$

We also have $u^3 = e$, $w^2 = e$, $uw = wu^2$ and these three equations are sufficient, in fact, to tell us all about this particular group.

11.2. The abstract definition of a group

It is convenient to repeat the relevant laws of chapter 10 in order to give a formal definition of a group.

Abstract definition of a group

A set S of elements forms a group if to any two elements x and y of S taken in a particular order there is associated a unique third element of S, called their product and denoted by xy, which satisfies the following laws:

M2 For any three elements x, y and z, $(xy)z = x(yz)$.

M3 There is an element called the neutral element and denoted by e which has the property that $xe = ex = x$ for all elements x.

M4 Corresponding to each element x there is an element x^{-1} called the inverse of x which has the property that

$$xx^{-1} = x^{-1}x = e.$$

Thus given a set and a product we need five things to make it a group.

(*a*) The product must be unique. This is usually implicit in its definition.

(*b*) The product must be in the set. This again is usually obvious from the definition.

(*c*) The Associative Law must hold. This is nearly always obvious, possibly needing a little thought.

(*d*) There must be a neutral element. It is usually clear which element this must be and it can then easily be tested.

(*e*) There must be an inverse to *every* element. This again is not often difficult to identify.

It seems from the above that it is usually a straightforward matter to test whether a structure is a group or not, and this is generally the case.

We give a few examples of structures which are *not* groups, in order to show a few of the pitfalls.

Example 1. All positive integers under multiplication. There are no inverses.

Example 2. All positive integers together with their reciprocals, under multiplication.

The product is not always within the set, e.g. $3 \times \frac{1}{4} = \frac{3}{4}$ and is not in the set.

Example 3. All complex numbers of the form $1+xi$ under addition.

Again the 'product' is not always in the set. (In fact it is never in.)

Example 4. All real numbers under multiplication.

0 has no inverse. If we except 0 then we do have a group.

Example 5. All transformations of a triangle which turn it over (possibly rotating as well).

There is no neutral element, nor is the product ever in the set.

11.3. Elementary consequences of the definition

Theorem 11.3.1. *The Associative Law extends to the product of more than three elements, i.e. the product $x_1 x_2 x_3 \ldots x_n$ is independent of the order in which we perform the multiplications of pairs, provided that we keep the x's in the same relative position.*

This follows by repeated application of M2. It is possible to give a formal proof by induction, but the result is so obvious that to do so would raise unnecessary complications.

An example for a particular case, that $(xy)(zt) = x((yz)t)$, is

$$\begin{aligned}(xy)(zt) &= x(y(zt)) \\ &= x((yz)t).\end{aligned}$$

Theorem 11.3.2. *The uniqueness of the neutral element.*

There cannot be two elements each having the property of e in M3.

The proof is exactly as in theorem 10.6.1.

Theorem 11.3.3. *The uniqueness of the inverse.*

The statement and proof are as in theorem 10.6.2.

Theorem 11.3.4. *The inverse of e is e.*

$ee = e$ by M3 and so e is its own inverse.

Theorem 11.3.5. $(x^{-1})^{-1} = x$.

The proof is as in theorem 10.6.4.

Theorem 11.3.6. *The Cancellation Law.*

$$xy = xz \Rightarrow y = z,$$
$$yx = zx \Rightarrow y = z.$$

The proof is as in theorem 10.6.5, except that we need make no exceptions, since there is no '0' in a group with a product structure and *every* element has an inverse.

Theorem 11.3.7. *The equation $ax = b$ has the unique solution $x = a^{-1}b$. The equation $xa = b$ has the unique solution $x = ba^{-1}$.*

The proof is as in theorem 10.6.6, again with no exceptions.

11.4. The order of a group

We see from the examples we have met so far that, while some groups have infinitely many elements (for example, the real numbers under addition, or the group of two-dimensional vectors under addition) others have only a finite number. In the latter case the number of elements is called the *order* of the group, and we say that the group is of finite order. In the case of infinite groups we say that they have infinite order, the order being the cardinal number of the set (see p. 163).

Thus the group of residues modulo 5 under multiplication has 4 elements and so is of order 4, as is the group of residues modulo 4 under addition. The transformation group of the triangle has order 6.

It is possible to have a group consisting of just one element e with $ee = e$. This has order 1 and is sometimes known as the trivial group.

It might be expected that nearly all important groups were infinite, since the sets used in elementary algebra (such as the real or complex numbers or the integers) are. This is not the case. While many infinite groups are extremely useful there is a large number of important finite ones, and these in fact are usually more interesting as groups. The reason is, roughly speaking, that in a finite group the structure must, after a certain stage, turn over upon itself and become intertwined; it cannot continue indefinitely along a straight course. For example, in the group of transformations of a triangle on p. 197 we found that although u^2 was a new element, $u^3 = e$. Also $w^2 = e$ and, furthermore, we had $uw = wu^2$. It is these relations that give the group its structure, and the structure is surprisingly interesting even when the order is as small as in this case. Such intertwinings and foldings back may occur of course in infinite groups, but it is usually in the finite case that they are exhibited to the highest degree. Thus a great deal of the interest and usefulness of the subject lies in finite groups, so much so that many works deal with these almost exclusively.

11.5. Abelian groups

In the definition of a group in §11.2 we did not assume the truth of the Commutative Law M 1. The reason was that it is not essential to much of our algebra and we wish in fact to work with many structures within which it does not hold. As an example we remember that the transformation group of the triangle is not commutative.

On the other hand many groups *are* commutative. Such are called *Abelian groups* (after the Norwegian mathematician N. H. Abel (1802–29) who anticipated some of the later work in group theory). They have of course properties which are not possessed by non-commutative groups, on the whole they are easier to work with, but in many ways they are not as interesting as the others.

It is usual, though not universal, to use the addition notation for Abelian groups. (Some writers use this for the non-commutative case also, so it is wise to be familiar with both notations!) We then have the four laws A 1–4 of chapter 10 with the corresponding elementary theorems. The neutral element becomes the zero and inverses are replaced by negatives. In many ways this notation is more familiar to handle, but there is a tendency when using it to assume commutativity: this of course does not matter when it is reserved for the Abelian case.

We will deal hardly at all with the theory of Abelian groups in this book, though many individual examples will be met with. Thus the addition notation will practically never be used, although we will meet with it in volume 2 when dealing with structures which possess two basic operations.

Even when the whole group is non-commutative certain pairs of elements may, of course, commute. Thus as $ex = xe = x$ the neutral element commutes with all others. Also any element commutes with its inverse by M4, and with all its powers, and other pairs may do so as well.

11.6. The multiplication table

For finite groups it is possible to write out all possible products of two elements in the form of a table. Suppose we have n elements. Then we take a table with n rows and n columns and place each element at the head of one row and one column, usually taking them in the same order for columns as for rows. In the space of the table which is the intersection of the row headed by x and the column headed by y we place the element xy. The table so formed has several properties and is particularly useful for groups of small order.

We give in figure 42 the multiplication table for the group of residues modulo 5 under multiplication.

	1	2	3	4
1	1	2	3	4
2	2	4	1	3
3	3	1	4	2
4	4	3	2	1

Fig. 42

We notice first that for an Abelian group the table is symmetrical about the leading diagonal—this would not be so for a non-commutative group.

If we put the neutral element against our first row and column then these merely give the elements in our chosen order.

The position of the neutral element in the other rows or columns gives us the inverses. Thus in the above the inverse of 2 is 3, that of 3 is 2, while 4 is its own inverse.

We see that all our postulates for a group are exhibited in the table except for the Associative Law, which it is difficult to verify from it.

In the above example every row or column contains each element once and only once, in some order. This is a general property as we see from the theorem below.

Theorem 11.6.1. *In any multiplication table for a finite group each element occurs exactly once in any row or column.*

Consider the row headed by the element a. The elements of this row are the elements ax, as x varies over the group. But by theorem 11.3.7 the equation $ax = b$ has a unique solution. Hence the element b occurs just once in the row. But b is any element, so the theorem follows for rows.

Similarly, since $xa = b$ has a unique solution, the theorem is true for columns.

Isomorphic groups

Suppose we form the multiplication table for the group of residues modulo 4 under addition.

	0	1	2	3
0	0	1	2	3
1	1	2	3	0
2	2	3	0	1
3	3	0	1	2

Fig. 43

At first sight this does not seem to bear much resemblance to the previous table of this section. Suppose, however, that in the first table we give the labels e, a, c and b to the elements 1, 2, 3, 4 in that order, and interchange the order of the rows and columns slightly to e, a, b, c. (It clearly does not matter in which order we take our elements in a group, except possibly that e should come first. In the case in question there is nothing sacrosanct in the ordering 1, 2, 3, 4: although we may think of 4 as greater than, say, 2, it is less than 7, which is another equally good member of the same residue class.) We obtain the table in figure 44.

We now see that this is precisely the same table as that above, with e, a, b, c replacing 0, 1, 2, 3.

	e (1)	a (2)	b (4)	c (3)
e	e	a	b	c
a	a	b	c	e
b	b	c	e	a
c	c	e	a	b

Fig. 44

Now the properties of groups depend only on the product which is defined in the set and which makes the set into a group —this product is the only aspect of the elements in which we are interested when we are investigating the set *as a group*. Hence from the group point of view the residues modulo 5 under multiplication behave in exactly the same way as the residues modulo 4 under addition, provided we pair 1, 2, 4, 3 of the first against 0, 1, 2, 3 of the second in that order. Any group property of elements of the first has its exact analogue for the corresponding elements of the second. For example, 4^3 ($= 4.4.4$) $= 4$ in the first group, and $2+2+2$ ($= 2.2.2$ in group product language) $= 2$ in the second. Again, $2x = 3$ has the root 4 in the first group, while $1x = 3$ (meaning $1+x = 3$ in ordinary notation) has the root 2 in the second.

Two such groups having exactly the same products under a suitable pairing off of elements are called *isomorphic* groups, the pairing off being known as an *isomorphism*. From the abstract point of view they are identical, and we sometimes speak of them as being the same group. The reader should refer back to §9.5 and note that, for groups, an isomorphism is a 1-1 correspondence that does not alter the product structure of the group.

There may be more than one isomorphism between two groups, i.e. more than one possible pairing off to make them isomorphic. Thus above we paired 1, 2, 4, 3 against 0, 1, 2, 3, but we could equally well have taken 1, 3, 4, 2 against 0, 1, 2, 3.

Isomorphism is not restricted to finite groups. Consider the set of integers, positive, negative and zero. These form a group, under addition as may be easily verified. The set of even integers also forms a group under addition, and these two groups are isomorphic under the isomorphism which relates n of the first to the even integer $2n$ which is a member of the second. (If the 'product' of m and n of the first is p we have $m+n = p$. Thus $2m+2n = 2p$ and so the 'product' of the corresponding elements is the element corresponding to their product.)

The symbol for 'is isomorphic to' is $\cong$, though we will often prefer to write it out in full, as this makes for easier reading and understanding.

Abstract groups

As we have already said the only thing that we need know about a given group is the product of any pair of elements. The group of residues modulo 5 under multiplication is described fully (*qua* group) in the table on p. 206, where the elements are denoted abstractly by letters. We have seen that the same table gives the group of residues modulo 4 under addition, and it will turn up again in the next chapter. Such a set of abstract symbols, combining in a given way, is known as an *abstract group*, and it is with these that we work when investigating group properties. This is an example of our familiar mathematical idea of abstraction. We isolate certain aspects of our set (in this case the group aspect), work with a model which exhibits these aspects without the extraneous properties possessed by the original, and then apply our results to our given set.

Two abstract groups may of course be isomorphic, but in this case it is usually merely a change in notation (thus in our example we could work with e, x, y, z instead of e, a, b, c), and two such are usually thought of as being the same.

The Vierergruppe

An important abstract group of order 4, which is not isomorphic to those above, is the *Vierergruppe*, or *four-group*. We will meet examples of it in the next chapter. Its multiplication table is given in figure 45 (overleaf).

	e	a	b	c
e	e	a	b	c
a	a	e	c	b
b	b	c	e	a
c	c	b	a	e

Fig. 45

11.7. The inverse of a product

In the set of vectors, let us consider a vector as a displacement and suppose that we have two vectors $\mathbf{x} = \mathbf{AB}$ and $\mathbf{y} = \mathbf{BC}$. Then the sum of $\mathbf{x}$ and $\mathbf{y}$ (i.e. the 'product' when we consider vectors as forming a group) is $\mathbf{AC}$.

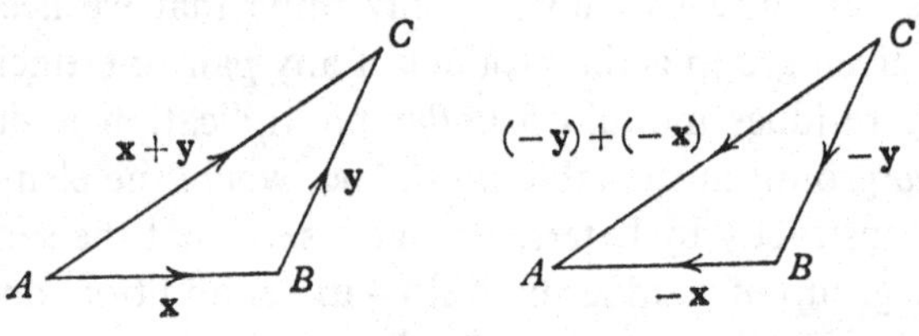

Fig. 46

The inverses of $\mathbf{x}$, $\mathbf{y}$, $\mathbf{x}+\mathbf{y}$ are $\mathbf{BA}$, $\mathbf{CB}$, $\mathbf{CA}$, respectively. The obvious way of writing $\mathbf{CA}$ in terms of the others is as $\mathbf{CB}+\mathbf{BA}$, *in that order*. Thus the inverse of $\mathbf{x}+\mathbf{y}$ is the inverse of $\mathbf{y}$ plus the inverse of $\mathbf{x}$, written naturally in this order, with $\mathbf{y}$ now replacing $\mathbf{x}$ as the first vector. Of course, since vector addition is commutative, this also equals $-\mathbf{x}-\mathbf{y}$, but the former order is the natural one and the latter merely a consequence of the commutativity.

In the transformation group of a triangle we have the same state of affairs. The inverse of a transformation is obtained by reversing the transformation—the inverse of a clockwise rotation is the same rotation but anticlockwise. If a transformation changes the position from the position A to the position B, then its inverse changes it from B to A. Thus, as in the case of displacements, to reverse the product of two transformations x and y, we must first apply the reverse (or inverse) of y, and *then* the inverse of x. In other words, the inverse of xy is $y^{-1}x^{-1}$, the order of x and y being changed. In this case the group is not commutative, and $y^{-1}x^{-1}$ is not necessarily the same as $x^{-1}y^{-1}$. Figure 47 shows how this works in a particular example.

This result is true generally, although it is significant only in the case of a non-Abelian group. The proof is given below, and is quite simple, the idea being that we work with the elements from the centre outwards, cancelling them in pairs.

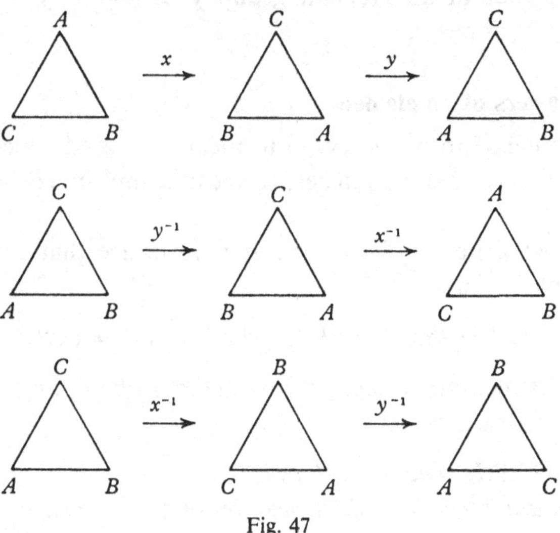

Fig. 47

Theorem 11.7.1. $(xy)^{-1} = y^{-1}x^{-1}$.

For
$$\begin{aligned}(y^{-1}x^{-1})(xy) &= y^{-1}(x^{-1}x)y \\ &= y^{-1}ey \\ &= y^{-1}y \\ &= e.\end{aligned}$$

Also
$$\begin{aligned}(xy)(y^{-1}x^{-1}) &= x(yy^{-1})x^{-1} \\ &= xex^{-1} \\ &= xx^{-1} \\ &= e.\end{aligned}$$

As an extension we have that $(xyz)^{-1} = z^{-1}y^{-1}x^{-1}$.

For
$$(xyz)^{-1} = (x(yz))^{-1}$$
$$= (yz)^{-1}x^{-1}$$
$$= z^{-1}y^{-1}x^{-1}.$$

Similarly for any number of elements, by induction.

The above theorem is easy to prove, but the result is extremely important, being rather surprising and a common source of error. To find the inverse of a product of two or more elements we must take the product of the inverses *in the reverse order*.

Notice that in an Abelian group $y^{-1}x^{-1} = x^{-1}y^{-1}$ and the difficulty disappears.

11.8. Powers of an element

We define x^2 to mean xx, x^3 to mean xxx, and generally x^r, where r is any positive integer, to mean x multiplied by itself r times.

We now write x^{-r} for $(x^{-1})^r$, and we notice that this is the inverse of x^r, for

$$(x^r)^{-1} = (xx \dots x)^{-1} = x^{-1}x^{-1} \dots x^{-1} = (x^{-1})^r.$$

As in elementary algebra we define x^0 to be e, and we then have the index laws.

Theorem 11.8.1. *The Index Laws.*

If m, n are integers, positive negative or zero, we have

$$\text{(i) } x^m x^n = x^{m+n}; \quad \text{(ii) } (x^m)^n = x^{mn}.$$

The proof is obvious from the definition of x^r for m and n positive, and we can prove it generally by considering all possible cases. None of these requires more than a moment's thought, and we will show the argument in one case only, when m is positive and n is negative and equals $-k$, say.

(i) $x^m x^n = x^m x^{-k} = x^m (x^{-1})^k$.

We now pair the x's with the (x^{-1})'s, starting from the middle, and we see that each pair becomes e and in effect cancels out. Thus if $m > k$ we are left with $(m-k)$ x's, while if $k > m$ we have $(k-m)$ (x^{-1})'s, i.e. in either case

$$x^m x^n = x^{m-k} = x^{m+n}.$$

(If $m = k$ we have $x^m x^n = e$ and of course $x^{m+n} = x^0 = e$.)

(ii) $(x^m)^n = (x^m)^{-k}$

$= ((x^m)^{-1})^k$ by definition of negative powers

$= ((x^{-1})^m)^k$ by the rule for the inverse of a product

$= (x^{-1})^{mk}$ by the Index Law for positive m and n

$= x^{-mk}$ by definition of negative powers

$= x^{mn}$.

It is important to notice that powers of the same element always commute, since $x^m x^n = x^n x^m = x^{m+n}$.

Theorem 11.8.2. *If two elements commute, so do their powers, i.e. If $xy = yx$, $x^m y^n = y^n x^m$.*

We can understand this best by proving a particular example.

$$x^2y^3 = xxyyy = xyxyy = yxxyy = yxyxy$$
$$= yyxxy = yyxyx = yyyxx$$
$$= y^3x^2.$$

The proof is similar for any positive m and n.

For m positive and n negative we notice that since $xy = yx$, $y^{-1}x = xy^{-1}$, by multiplying both sides on the left and right by y^{-1}. Hence $x^m(y^{-1})^k = (y^{-1})^k x^m$ where $k = -n$, i.e. $x^m y^n = y^n x^m$. Similarly for m negative, n positive, and for both negative.

The order of an element

Suppose we take the element 2 of the group of residues modulo 5 under multiplication, and consider its powers. $2^0 = 1$, $2^1 = 2$, $2^2 = 4$, $2^3 = 8 = 3$ and $2^4 = 16 = 1 = 2^0$. Thus the first 4 powers are all different, but after that the powers repeat, for $2^4 = 2^0$, $2^5 = 2^1$, $2^6 = 2^2$, etc. Since there are only four different elements in the group, it is clear that powers must repeat, and we see that before this repetition we work through all the four different elements. This is not always the case, as, for example, the powers of the element 4 in the same group are $4^0 = 1$, $4^1 = 4$, but $4^2 = 16 = 1$, and then of course we have repetitions as before. Thus the only possible results for powers of 4 are the two elements 1 and 4. We may similarly verify that 3^n takes all possible values, but 1^n is always 1, taking the single value only.

Again, in the transformation group of the triangle, $u^0 = e$, $u^1 = u$, $u^2 = v$ but $u^3 = e$ again and from then on we repeat. Similarly, the powers of v are only e, v, u and then e again, while the squares of w, x and y are all e and so the powers of any of these three elements take just two values, the element itself and e.

The number of possible values of powers of an element x is called the *order* of x. In a finite group it must be a finite number not greater than the number of elements in the group, i.e. the order of the group. But it may be less than this as we see in the case of 4 in the residue group above, and indeed the orders of all elements may be less than that of the group, as in the transformation group of the triangle. The first element to be repeated, as we list the positive powers in order, is always, as we have seen, the neutral element e, and after this all elements are repeated indefinitely.

The simple results concerning order are proved below, for any group. Note that we define order as the lowest positive power to be equal to e, and theorem 11.8.3 proves that all lower positive powers are distinct.

Let us consider the set of elements e $(= x^0)$, x, x^2, x^3, ..., where x is some element of our group. In an infinite group it is possible for all these to be different (for example, $x = 2$ in the group of real numbers except 0 under multiplication) but in a group of finite order we must sooner or later have two which are the same element. Suppose $x^m = x^n$ where $m < n$. Then $x^{n-m} = x^m x^{-m} = e$.

Hence in a finite group some positive power of any element must equal the neutral element. If n is the least positive integer such that $x^n = e$ we say that n is the *order* of the element x.

Theorem 11.8.3. *If x has order n the n elements $e, x, x^2, \ldots, x^{n-1}$ are all distinct.*

Suppose $x^k = x^l$, where $0 \leqslant k < l \leqslant n-1$. Then $x^{l-k} = e$. But $0 < l-k < n$, and hence our supposition that $x^k = x^l$ is impossible by the definition of order of x, i.e. the first n powers of x are distinct.

It is obvious that $x^{n+1} = x$, $x^{n+2} = x^2$ and so on. In general $x^{rn+s} = x^s$, and $x^m = e \Leftrightarrow m$ is a multiple of n, possibly the zero multiple.

We see also that $x^{-1} = x^n x^{-1} = x^{n-1}$, $x^{-2} = x^{n-2}$, etc., and that $x^{-n} = e$, so that the order of x is the same as the order of x^{-1}.

The order of e is 1, and of course e is the only element of order 1. Note that if x has order 2 we have $x = x^{-1}$ and conversely.

The order of any element is not greater than the order of the group, since by theorem 11.8.3 we have n *distinct* elements if x has order n.

In an infinite group there may be no n such that $x^n = e$, in which case all elements $e, x^{\pm 1}, x^{\pm 2}, \ldots$ are distinct. In this case we say that x has infinite order. There may, however, be elements of finite order even in an infinite group. For instance, in the group of real numbers excluding 0 under multiplication the element -1 has order 2, and of course the neutral element always has order 1.

As a further example of orders let us consider the group of residues modulo 4 under addition. The order of 0 is of course 1, those of 1 and 3 are both 4 (which equals the order of the group and is the maximum possible), while 2 has order 2, since $2+2 = 0$ modulo 4.

It was simple to discover the orders of the elements of the group of residues modulo 5 under multiplication, but we remember that this group is isomorphic to the above group of residues modulo 4, and hence that corresponding elements must have the same order. Thus in the residues modulo 5 the order of 1 is 1, that of both 2 and 3 is 4, while 4 has order 2, as before.

11.9. Latin squares

In any multiplication table of a finite group we notice that no element occurs twice in any row or column since, by the Cancellation Law, ab cannot equal ac unless $b = c$, and similarly for ba and ca. A square of n^2 compartments in which n elements are placed in such a way that this property is true is called a 'Latin square'.

Latin squares are much used in sampling. For instance, if we wish to test n different types of seed in agricultural research in such a way that we lessen the differences introduced by the nature of the testing ground we may sow n^2 plots in the form of a Latin square.

Although the multiplication table of a group is always a Latin square, all such squares cannot be represented in this way, since there is no guarantee that the Associative Law holds. In fact there are many more than those obtained by group theory, but the use of the latter will at least produce some squares.

Worked exercises

1. Is the set of all rational numbers a group under addition?

If p/q and r/s are any two rational numbers, $(p/q)+(r/s) = (ps+qr)/qs$ and is rational, i.e. is in our set. The sum is of course unique and addition is associative.

The neutral element is 0 and the inverse of p/q is $-(p/q)$. Hence the set of rationals does form a group under addition.

2. Is the set of positive rational numbers a group under the definition of product of x and y as x/y?

The 'product' is rational and so in the set—it is also unique. $x.1 = x$, but $1.x = 1/x$ and so 1 is not a neutral element, and we can easily see that there is no neutral element. Hence there cannot be an inverse defined (but note that $(p/q)(p/q) = 1$).

Also the product is not associative, since

$$(xy)z = \left(\frac{x}{y}\right)\Big/ z = \frac{x}{yz} \quad \text{while} \quad x(yz) = x\Big/\left(\frac{y}{z}\right) = \frac{xz}{y}.$$

Hence we do not have a group.

3. Suppose a, b are elements of a group and that $a^6 = e$, $b^4 = e$, $ab = ba^3$. Prove (i) $a^2b = ba$; (ii) $ab^3 = b^3a^2$.

(i) $a^2b = a.ab = aba^3 = ba^3a^3 = ba^6 = ba$, since $a^5 = e$.

(ii) $ab^3 = abb^2 = ba^3b^2 = baa^2b^2 = ba.a^2b.b = ba.ba.b$ by (i)

$= b.ab.ab = b.ba^3.ab = b^2a^4b = b^2a^2.a^2b$

$= b^2a^2.ba$ by (i)

$= b^2.a^2b.a$

$= b^3a^2$ by (i) again.

The method in the above is gradually to work the b's to the left of the expression and the a's to the right. This is simple in (i) since we can move one a at a time without introducing any more awkward b's, but in (ii) the process is longer. There is no infallible rule for such manipulations, and in many cases the process may be exceedingly difficult. (ii) may be proved more easily by multiplying both sides of (i) by b^3 both in front and behind and using the fact that $b^4 = e$.

4. Suppose a, b are elements of a group and $a^2 = e$, $b^6 = e$, $ab = b^4a$. Find the order of ab, and express its inverse in each of the forms $a^m b^n$ and $b^m a^n$.

$(ab)^2 = b^4a.ab = b^5$, since $a^2 = e$. Hence $(ab)^3 = ab^6 = a$ and so $(ab)^6 = a^2 = e$. We see that $(ab)^4 = a.ab = b$ and $(ab)^5 = ab^2$.

Thus $(ab)^6 = e$ but no smaller power of ab is e, and so ab has order 6.

Since ab has order 6, $(ab)^{-1} = (ab)^5 = ab^2$ as above. Also $ab^2 = b^4ab = b^8a = b^2a$ for the other form.

Note that we can say more about these elements. $(ab)^6 = e$. But $(ab)^2 = b^5$ and so $b^{15} = e$, i.e. $b^3 = e$. Thus the relations simplify to $a^2 = e$, $b^3 = e$, $ab = ba$.

Exercises 11A

Which of the sets **1–13** are groups under the given operations? If the set is a group give the neutral element and the inverses; if it is not state which laws are not satisfied.

1. The even integers (positive, negative or zero) under addition.

2. The positive integers, including zero, under addition.

3. All integers, positive, negative or zero, divisible by 7, under addition.

4. All integers, positive or negative, which include the digit 5, including zero, under addition.

5. All numbers of the form 2^n where n is an integer, positive, negative or zero, under multiplication.

6. All rational numbers under multiplication.

7. All rational numbers excluding zero under multiplication.

8. All two-dimensional vectors of the form $(x, \lambda x)$ where λ is a fixed scalar under vector addition.

9. All vectors of the form either $(0, y)$ or $(x, 0)$ under vector addition.

10. Residues modulo 8 under addition.

11. Residues modulo 8, excluding 0, under multiplication.

12. The elements 0 and 1 with product defined as $0.0 = 0$, $0.1 = 0$, $1.0 = 0$, $1.1 = 1$.

13. The elements 0 and 1 with product defined as $0.0 = 0$, $0.1 = 1$, $1.0 = 1$, $1.1 = 0$.

In the transformation group of the triangle, with the notation of the text, prove the identities **14–16**:

14. $u^2w = wu$. **15.** $uwu = w$. **16.** $(wu)^2 = e$.

17. What are the orders of the 6 elements of the transformation group of the triangle?

18. What are the inverses of the elements of the transformation group of the triangle? Illustrate the answer by diagrams.

Construct the multiplication table for the groups **19–22**, and give the orders and inverses of each element.

19. The group of exercise **13**.

20. The group of residues modulo 5 under addition.

21. The group of residues modulo 6 under addition.

22. The group of residues modulo 7 under multiplication (except 0).

23. Show that the groups in **21** and **22** above are isomorphic and give an isomorphism.

24. Give an example of a group in which every element except e has order 2.

25. If $b = x^{-1}ax$ show that $b^2 = x^{-1}a^2x$ and hence show that a and b have the same order.

26. Prove that if x has order n, the order of any power of x is not greater than n. Prove also that if n is prime the order of any power of x (other than a power which is a multiple of n) is precisely n.

27. Prove that the elements xy and yx have the same order.

Exercises 11B

1. Suppose that a and b are elements of a certain group and that $a^3 = e$, $b^4 = e$, $ba = ab^3$. Prove the identities

(i) $b^2a = ab^2$;
(ii) $b^3a = ab$;
(iii) $bab = a$;
(iv) $b^{-1}a^{-1}b^{-1} = a^2$.

What are the orders of ba, ab, ab^2?

2. Suppose that a and b are elements of a group such that $a^2 = e$, $b^5 = e$, $ab = b^4a$.

(i) Express ab^2, ab^3, ab^4 in the form b^ma^n.
(ii) What are the orders of ab, ba, ab^2?
(iii) Express the inverses of ab, ba, ab^3 in the form a^mb^n and also in the form b^ma^n.

3. Assuming in exercises **1** and **2** above that every element of the group is expressible in terms of a and b, what are the orders of the groups concerned?

Explain your answers and give a list of the elements in terms of a and b.

4. Suppose we have a set S in which a unique product is defined which is associative. Suppose also that $\exists$ a left neutral element (i.e. an element e such that $ex = x \; \forall \, x \in S$) and a left inverse of every element x (i.e. an element y such that $yx = e$). Prove that e is also a right neutral element, y is a right inverse and that S is a group under the given product.

5. Suppose we have a *finite* set S with an associative product defined. Suppose also that the Cancellation Law (theorem 11.3.6) holds. Prove that M3 and M4 are true and hence that S is a group.

6. Prove that if x, y, xy are all of order 2 then $xy = yx$. Hence show that if every element of a group is of order 2 (except e) the group is Abelian.

7. In a group of even order prove that there is at least one element of order 2.

12

EXAMPLES OF GROUPS

In this chapter we give a large selection of groups which occur in practice or in other branches of mathematics. They include many which are important from the theoretical point of view and we will draw upon these examples in the next chapter and in volume 2; indeed many of them will appear again as other structures in that volume. The verification that a certain set is a group will often be left to the reader: it is not usually difficult.

12.1. Examples from sets of numbers

Most of our ordinary sets of all numbers form groups, under addition or multiplication or both. We give also some interesting subsets which are groups in their own right.

The real numbers

These form a group under addition and also, if we exclude 0, under multiplication.

The complex numbers

These likewise are a group under addition and, if we exclude 0, under multiplication.

The Quaternions

An Abelian group under addition, but (excepting 0) a non-commutative one under multiplication.

The rationals

Since the sum and product of two rationals are themselves rational we see that the rationals form groups under addition and (excluding 0) multiplication.

The integers under addition

If we take the set of all integers, positive, negative and zero, this is clearly a group under addition. It is an extremely important example, known as the *group of integers* or the *infinite cyclic group*. The reason for the latter name will be seen in §12.2.

The Gaussian integers

The complex numbers $a+bi$ where a and b are integers, are called *Gaussian Integers*. They form a group under addition.

The set of multiples of r

We consider all integers, positive negative or zero, which are multiples of a fixed integer r. Under addition these form a group. All such groups are isomorphic to the group of integers, the isomorphism being given by the correspondence of a in the group of integers with ar in the other set. A particular case is the group of even integers.

We could take multiples of *any* number, not necessarily an integer.

The positive numbers

The positive real numbers and the positive rationals both form groups under multiplication.

The set $\{x^n\}$ under multiplication

If n is an integer, positive negative or zero, and x is any number (real or complex) then the set $\{x^n\}$ is a group under multiplication, since $x^m . x^n = x^{m+n}$ and is in the set.

All such groups are isomorphic to the group of integers (unless x is 0 or a root of unity), under the isomorphism n corresponding to x^n, since

$$x^m . x^n = x^{m+n}$$

and corresponds to $(m+n)$, the group product of m and n in the group of integers.

The set $a+b\sqrt{n}$ under addition

Consider all numbers of the form $a+b\sqrt{n}$, where a and b are integers and n is some integer which is not a perfect square. Then these form a group under addition. For

$$(a_1+b_1\sqrt{n})+(a_2+b_2\sqrt{n}) = (a_1+a_2)+(b_1+b_2)\sqrt{n}$$

and is in the set, addition is associative, the neutral element is $0+0\sqrt{n}$ and the inverse of $a+b\sqrt{n}$ is $-a-b\sqrt{n}$.

We notice that all these groups are isomorphic to each other and to the group of Gaussian integers. They are also isomorphic to the group of two-dimensional vectors with integral co-ordinates. The $\sqrt{n}$, in fact, acts merely as an indeterminate symbol; its value does not affect our structure, and we could in fact take any real number for n so long as it were not a perfect square.

If n is a perfect square we still have a group but it is no longer isomorphic to the others. It is, of course, merely the ordinary group of integers.

If a and b may take any *rational* values and n is a rational number which is not a perfect square (or an irrational number) we obtain another group, all such again being isomorphic.

The set $a+b\sqrt{n}$ under multiplication

If a and b are rational and not both zero and n is a rational number, not a perfect square, then the set $a+b\sqrt{n}$ (excluding 0) form a group under multiplication. $(a_1+b_2\sqrt{n}).(a_2+b_2\sqrt{n}) = (a_1a_2+nb_1b_2)+(a_1b_2+a_2b_1)\sqrt{n}$ and is in the set. Product is of course associative, and $1+0\sqrt{n}$ is a neutral element.

To find an inverse, we notice that

$$\frac{1}{a+b\sqrt{n}} = \frac{a-b\sqrt{n}}{a^2-nb^2} = \frac{a}{a^2-nb^2} - \frac{b}{a^2-nb^2}\sqrt{n}.$$

Now since n is not a perfect square $a^2-nb^2 \neq 0$ and so the above, being of the form $c+d\sqrt{n}$ with c and d rational, is in the set. Hence the set *is* a group.

It is important to notice that all such groups are *not* isomorphic, since the number n enters directly into the product and is not merely a symbol. Note also that if a, b and n are restricted to the integers we no longer have a group since the inverse is not now in the set.

The four numbers $1, -1, i, -i$ under multiplication

If we construct the multiplication table we find that this group is isomorphic to the group of residues modulo 4 under addition, one isomorphism being $1, i, -1, -i$ corresponding to 0, 1, 2, 3, respectively.

Complex numbers of unit modulus

The product of two such numbers is in the set, the unity is 1, and the inverse is the complex conjugate.

12.2. Residue classes: the cyclic groups

Groups of residues under addition

If we consider the set of residue classes modulo n, where n is any given positive integer, we recall that in §6.2 we showed that it is possible to add and subtract two residues. To do this we take any representatives from the two classes, add or subtract these representatives and take the class containing our answer to be the required sum or difference: we showed that this is independent of the particular representatives chosen. It is convenient to denote the residues by their smallest positive members; our set then consists of $0, 1, 2, \ldots, (n-1)$ and to add or subtract we merely perform the ordinary process with these numbers and ignore any multiples of n.

We see that under addition this set forms a group, of order n. The neutral element is 0 and the inverse of r is $(n-r)$.

Let us consider the positive powers (in the group sense) of the element 1. They are $0, 1, 2, \ldots, (n-1)$ and the nth power is n, which equals 0, the neutral element. Thus the order of 1 is n and the first n powers of 1 are precisely the elements of the group. Thus if we denote 1 by x the n elements are just $e, x, x^2, \ldots, x^{n-1}$ with $x^n = e$. We call such a group a *cyclic group* and say that x is a *generator*, since all elements are found in terms of x.

Cyclic groups

A cyclic group is one which consists merely of powers of one of its elements, called a generator. If the generator x is of finite order n then the elements are $e, x, x^2, x^3, \ldots, x^{n-1}$ and of course $x^n = e$. All cyclic groups of order n are isomorphic, an isomorphism being obtained by making the generators correspond, and we have seen that an example is given by the group of residues modulo n under addition. We speak of *the* cyclic group of order n, and we see that this ensures the existence of at least one group of each finite order. (Often there is more than one, of course.)

The cyclic groups are Abelian. The cyclic group of order n is sometimes denoted by C_n.

From what we have already shown other examples of the cyclic group of order 4 are the group of residues modulo 5 under multiplication, and the group consisting of ± 1, $\pm i$ under multiplication.

Suppose the generator has infinite order. In this case the elements $e, x^{\pm 1}, x^{\pm 2}, \ldots$ are all different and the group consists of these. Again all such groups are isomorphic, and the group is the *infinite cyclic group*. We have already met it on p. 219, where we saw that it was isomorphic to the group of integers.

In the infinite cyclic group there are just 2 possible generators, x and x^{-1}, but in a finite one there may be several. Any element order n can be taken, since its first n powers are all distinct and therefore include the whole group.

Consider the cyclic group of order 6, formed by the powers $x^r, r = 0, \ldots, 5$. x itself is a generator and has order 6. The positive powers of x^2 are $e, x^2, x^4, e, \ldots$ and so x^2 has order 3 and cannot be a generator of the whole group. Similarly, x^3 has order 2, while x^4 has order 3 again. But the powers of x^5 are e, x^5, x^4, x^3, x^2, x and then e again, so that x^5 *is* a generator.

We saw that the only elements that arose as powers of x^2 or x^4 were x^0, x^2 and x^4, that is all the even powers of x. Similarly, the only ones that arose as powers of x^3 were x^0 and x^3, i.e. powers of x which are multiples of 3. Now 3 is the H.C.F. of 6 (the order of the group) and 3 (the power of x we are considering), while 2 is the H.C.F. of 6 and either 2 or 4.

The same is true generally. If we have the cyclic group of order n and consider the powers of x^r, where the H.C.F. of n and r is h say, the only elements that arise as powers of x^r are x^{kh}. Thus x^r is a generator if and only if $h = 1$, i.e. if n and r are co-prime. We proceed to prove this rigorously.

Theorem 12.2.1. *In the cyclic group of order n generator x the element x^r has order n, i.e. is a generator, if and only if r is prime to n.*

Suppose $(x^r)^k = e$. Then $x^{rk} = e$ and so rk must be a multiple of n. If r is prime to n this means that k must be a multiple of

n and so the first positive power of x^r to equal e is the nth. Hence x^r has order n. Conversely if r is not prime to n there is a value of k less than n (namely n/h where h is the H.C.F. of r and n) which makes rk a multiple of n, and so the order of x^r is less than n.

Theorem 12.2.2 *If r is not prime to n the order of x^r is l/r, where l is the* L.C.M. *of r and n.*

For the lowest k to make rk a multiple of n occurs when $rk = l$, and hence when $k = l/r$.

Groups of residues under multiplication

We remember from chapter 6 that we can always multiply residues but that division is not always possible. This means that if we try to form a group of residues modulo n under multiplication we will not always be successful. The product is always uniquely defined and in the set, and is associative, and we have a neutral element 1, but we do not always have an inverse. In fact by theorem 6.3.1 the residue r has an inverse if and only if r is prime to n and r^{-1} is also prime to n since $rr^{-1} \equiv 1 (\text{mod}\ n) = \lambda n + 1$ and so r^{-1} and n can have no common factor.

Thus if we take the set of residues prime to n each will have an inverse, and we still include the neutral element 1. Furthermore, if r and s are both prime to n so is rs, and so the product of any two elements is still in the set. Hence the residues prime to n form a group under multiplication.

If n is prime this group includes all elements except 0, i.e. its order is $(n-1)$. For the case $n = 5$ we have already investigated the group.

The order of the group, the number of numbers less than n prime to n, is denoted by $\phi(n)$ and is known as *Euler's function.*

These groups are interesting, though by no means so important as the cyclic groups, since their structure can be fairly complicated. For example, if we consider the powers of an element of a cyclic group they form a fairly obvious sequence, proceeding round the group as it were in an orderly fashion. But the powers of elements in the multiplication groups of residues behave in a much more haphazard fashion. As an example let us take the residues modulo 7. Since 7 is prime, the group is formed of the six residues 1, ..., 6. 1 is the neutral element and has order 1. The powers of 2 are equal respectively to 1, 2, 4, 1, 2, 4, 1, ... and so the order of 2 is 3. The powers of 3 are 1, 3, 2, 6, 4, 5, 1, ... and so 3 has order 6 and is in fact a generator. Similarly, we can see that the order of 4 is 3, that of 5 is 6 (so 5 is a generator) and that of 6 is 2. Thus 3 and 5 are generators, while the orders of the other elements are 3, 2 or 1.

Now consider the residues modulo 8. The only ones prime to 8 are 1, 3, 5, and 7, and these form our group. The order of 1 is of course 1, while since $3^2 = 9 = 1$, $5^2 = 25 = 1$, $7^2 = 49 = 1$ all the other elements have order 2. Thus there is no single generator in this case.

We can see then that the behaviour of these groups is by no means straightforward, and there is no theorem corresponding to theorem 12.2.1 above, nor any so comprehensive as theorem 12.2.2. We do however notice that in the group modulo 7 all orders are divisors of 6, the order of the group, while modulo 8 they are divisors of 4, the order of that group. This is true generally—the order of any element of any of these groups is always a factor of the order of the group. The proof, given below, is a good example of the use of Fermat's theorem and Euler's extension of it.

Theorem 12.2.3. *In a group of residues under multiplication, the order of any element is a factor of the order of the group.*

(i) In the group of residues modulo n, where n is prime, we have by Fermat's theorem (theorem 6.6.1) that if x is prime to n (this is the case for all elements of our group) $x^{n-1} = 1$ (mod n), and hence the order of x is a factor of $(n-1)$, the order of the group.

(ii) When n is not a prime we have by theorem 6.6.2 that $x^{\phi(n)} = 1$ (mod n) for all x prime to n, i.e. for all our elements. But $\phi(n)$ is the order of the group, and so the result follows.

The above theorem is a special case of a general result (theorem 13.7.2). Note that for the cyclic groups it is proved in theorem 12.2.2, since l/r is certainly a factor of n, for it is equal to n/h, where h is the H.C.F. of n and r.

12.3. Groups of polynomials

Polynomials under addition

All polynomials in a single variable x clearly form a group under addition, the neutral element being the polynomial 0 and the inverse of $P(x)$ being $-P(x)$.

In particular all polynomials of degree at most n form a group under addition, since the sum of two such is another. For this to be true always we must allow any coefficient or coefficients to be zero, including the coefficient of x^n. The neutral element is the polynomial 0, i.e.

$$0x^n + 0x^{n-1} + \dots + 0x + 0,$$

and as before the inverse of $P(x)$ is $-P(x)$.

We see that if we write the polynomial

$$a_n x^n + a_{n-1} x^{n-1} + \dots + a_0$$

(of degree n) as an ordered set of coefficients $(a_n, a_{n-1}, \dots, a_0)$ the sum of $(a_n, a_{n-1}, \dots, a_0)$ and $(b_n, b_{n-1}, \dots, b_0)$ is

$$(a_n + b_n, a_{n-1} + b_{n-1}, \dots, a_0 + b_0).$$

This is the same result that would be obtained if we were adding $(n+1)$-dimensional vectors, and so the group of polynomials of degree at most n under addition is isomorphic to the group of $(n+1)$-dimensional vectors.

In the above we may think of the coefficients as real numbers, rationals, complex numbers, or integers, or indeed as elements of any group (with the addition notation), since to form the sum of two polynomials we merely add the corresponding coefficients.

Polynomials under multiplication

If we multiply two polynomials we obtain another, usually of higher degree, provided our coefficients can be added and multiplied. We have a neutral element in the polynomial 1, but there is no inverse in general, although the constant polynomials do have inverses provided the coefficients have. Hence polynomials do not form a group under multiplication.

If instead of polynomials we think of the set of rational functions these, with the exception of 0, do form a group.

(Provided the coefficients can be added, subtracted and multiplied and are such that $ab = 0 \Rightarrow a = 0$ or $b = 0$.)

It appears that from the group point of view polynomials are not very interesting. Their importance lies in the fact that they can be multiplied as well as added: they form an important example of a *ring*, and will be dealt with as such in volume 2.

12.4. Vectors

n-dimensional vectors form a group under addition. As for multiplication we recall that there is no general definition of product of two vectors. In the two-dimensional case we can think of them as complex numbers, in which case the product structure is of course a group. For three-dimensions the scalar product produces a scalar and not a vector and so cannot give us a group, the product not being in the set. As for the vector product, we saw in theorem 10.7.1 that it was not associative. Hence this also does not produce a group.

The other algebraic operation, apart from addition, which may always be performed on vectors is that of multiplication by a scalar. Vectors therefore have more structure than that possessed by virtue of their being groups: they are an example of a *vector space* and some of their properties as such are investigated in volume 2.

12.5. Transformations of polygons: the dihedral groups

An important type of group is that consisting of all possible transformations of a figure. The transformation group of a triangle was dealt with in example 4 of §11.1; in this section we will introduce the transformation groups of general regular polygons, while in §12.7 solid figures will be dealt with.

We remember that we defined a transformation to be a movement of our figure which leaves it occupying the same position, two movements which have the same effect being taken to be the same. The product of two transformations is defined to be that which has the same effect as when we apply one transformation and follow it by the second. Under this definition the set of all possible transformations on any figure always forms a group; since product is associative, there is a neutral element (the identity transformation which leaves the figure unaltered) and there is an inverse to each transformation (that which has the reverse effect). Such groups give a measure and a description of the symmetry of our figure and provide a way of isolating the symmetry properties. They are particularly useful in the case of regular polygons, solids or figures of higher dimension: figures which are semi-regular may be identified with a regular one in this respect, while for an object with no symmetry the group of transformations consists merely of the trivial group $\{e\}$.

Transformations of a one-faced polygon

Consider a regular n-gon which can move in its plane, but which we imagine cannot be turned over (i.e. the two sides are taken to be different, coloured differently, for example). The

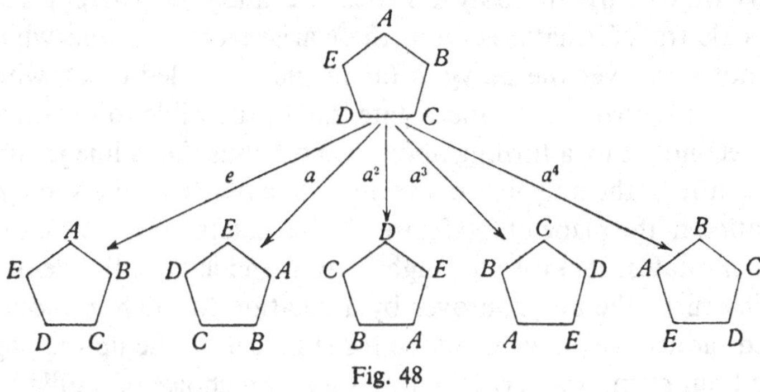

Fig. 48

only transformations are rotations clockwise or anticlockwise through multiples of $2\pi/n$ radians. Let us call that which rotates through $2\pi/n$ clockwise a. Then a^2 rotates through $2(2\pi/n)$ clockwise, a^3 through $3(2\pi/n)$, etc., while a^n rotates through 2π clockwise and hence leaves the polygon unchanged: i.e. $a^n = e$, the identity transformation. We see that the n elements $e, a, a^2, \ldots, a^{n-1}$ are the only different elements of our group; they take one particular vertex into all n in turn and this fixes the position of the polygon. We see also that $a^{n+1} = a$, $a^{n+2} = a^2$, etc., while $a^{-1} = a^{n-1}$, $a^{-2} = a^{n-2}$, etc. *Hence the group is the cyclic group of order n, a being a generator.*

The 5 elements for a regular pentagon are shown in figure 48.

Transformations of a circle

A circle may be rotated through any angle, positive or negative, taking clockwise as positive, but two angles which differ by 2π give the same transformation. The product of two rotations is obtained by adding their angles algebraically.

Hence the transformation group is isomorphic to the group of real numbers under addition, modulo 2π. We could change this modulus without altering the group (for example, we could work modulo 360, in degrees) and for convenience it could be taken as 1.

Transformations of a two-faced polygon

We take a regular n-gon whose two faces are identical, and we allow movements in three dimensions, so that transformations which turn the polygon over are allowed. There are $2n$ possible transformations. n of these arise from rotations which do not turn over the polygon (as in the one-sided case), while the other n turn it over and rotate it. It is possible to obtain all the second set by a turning about some radius (i.e. a line joining the centre of the polygon to a vertex) fixed *in space* followed by a rotation in the plane of the figure. Let us call the transformation which rotates clockwise through $2\pi/n$, a, and let b be the element which turns the polygon over by a rotation through π about a fixed radius, which we will take for simplicity to lie up the page in all our examples. (We see that there is a choice of n different elements for b: the group may be described in these many, at least, different ways. This is analogous to the case of the cyclic groups having a choice of generators.)

The n elements which do not turn over the polygon are, as in the one-sided case, $e, a, a^2, \ldots, a^{n-1}$ with $a^n = e$. The other n elements are $b, ba, ba^2, \ldots, ba^{n-1}$. We also have the relation that $b^2 = e$.

Theorem 12.5.1. *$ab = ba^{n-1}$ and, more generally, $a^r b = ba^{n-r}$.*

The fact that $ab = ba^{n-1}$ is seen from figure 49. Similarly, we can show the second part, as is done for $r = 3$ in figure 50.

Alternatively, we can show the second result from the first, by induction.

Suppose $\quad a^r b = ba^{n-r}.$

Then
$$a^{r+1}b = a.a^r b = a.ba^{n-r} = ba^{n-1}.a^{n-r}$$
$$= ba^{2n-r-1} = ba^{n-(r+1)}.$$

But the result is true for $r = 1$ and so it is generally true by induction.

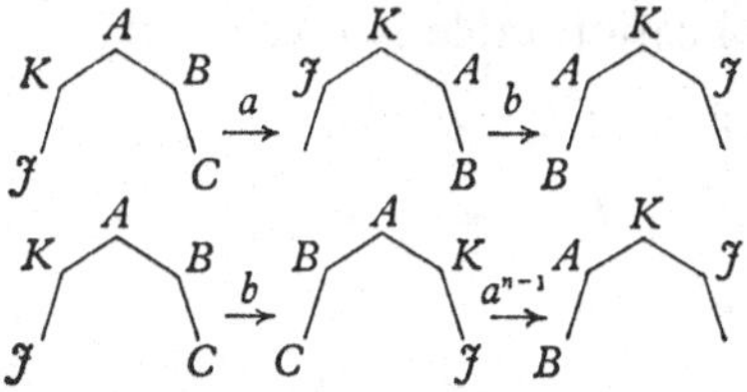

Fig. 49

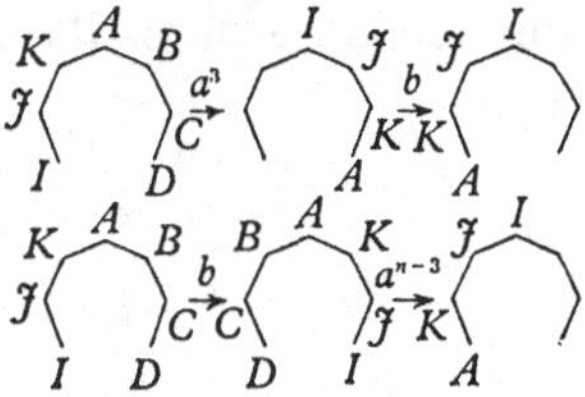

Fig. 50

The group is described completely by the three relations $a^n = e$, $b^2 = e$, $ab = ba^{n-1}$, where a and b are elements and b is not a power of a. The third relation enables us to put any product of a's and b's in the form ba^r: in other words turning over first and then rotating.

The group is known as the *dihedral group of order* $2n$. It is non-Abelian for $n \geqslant 3$, and we will sometimes write it as D_{2n}.

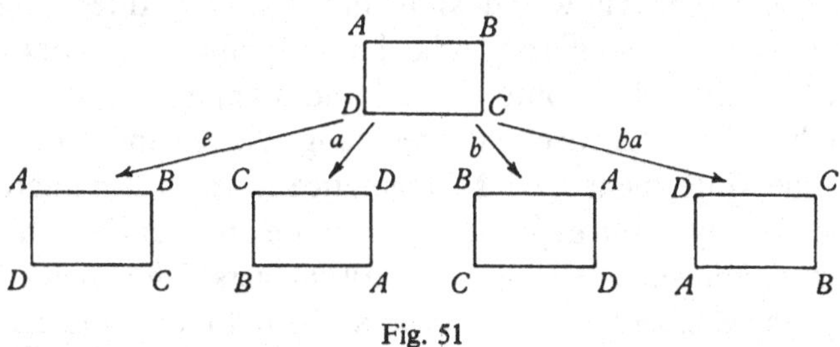

Fig. 51

The dihedral group of order 4

This is the transformation group of a line segment which may be turned over in two perpendicular directions. It is simpler to think of it as the transformation group of a rectangle in space (not a square—this gives the dihedral group order 8).

The four elements are given in figure 51. In this case our relations are $a^2 = e$, $b^2 = e$, $ab = ba$. The group is Abelian and when its multiplication table is constructed we notice that it is the Vierergruppe of p. 207 with ab written in place of c. Every element (except e) has order 2.

The dihedral group of order 6

This is the group of the equilateral triangle and was investigated in chapter 11. It is non-Abelian, and in fact is the smallest non-Abelian group. Its multiplication table is given here.

	e	a	a^2	b	ba	ba^2
e	e	a	a^2	b	ba	ba^2
a	a	a^2	e	ba^2	b	ba
a^2	a^2	e	a	ba	ba^2	b
b	b	ba	ba^2	e	a	a^2
ba	ba	ba^2	b	a^2	e	a
ba^2	ba^2	b	ba	a	a^2	e

Fig. 52

12.6. Permutations

If we have a set of distinct objects a *permutation* of them is a 1-1 mapping of the set onto itself. For example, suppose we have three vases, red, yellow and white, one in each of three different rooms. Then we can permute them by mapping the red one onto the white, the yellow onto the red and mapping the white one onto the yellow. Or we could interchange the red and yellow and leave the white unchanged. Mathematically we are not interested in the identity of the objects: we can represent them by symbols, the most convenient being ordinary integers. Thus if we have n objects we label them 1, 2, 3, ..., n. Note that the integers are here used merely as labels: they have no arithmetical meaning, but their use does express the fact that all the objects are distinguishable from one another. A permutation then maps the numbers (i.e. the objects) 1, ..., n onto themselves in a different order. If it maps the number i onto a_i we write it in the form

$$\begin{pmatrix} 1 & 2 & 3 & \dots & n \\ a_1 & a_2 & a_3 & \dots & a_n \end{pmatrix}.$$

Thus the permutation $\begin{pmatrix} 1 & 2 & 3 \\ 3 & 2 & 1 \end{pmatrix}$

means that we map 1 onto 3, 2 is left alone, and 3 is mapped onto 1.

We are not interested in the original order 1, 2, ..., n but merely in their images. Hence we may write the top row of the permutation in any order, so long as we make the corresponding changes to the bottom row. Thus we may write

$$\begin{pmatrix} 1 & 2 & 3 \\ 3 & 2 & 1 \end{pmatrix} \text{ as } \begin{pmatrix} 2 & 3 & 1 \\ 2 & 1 & 3 \end{pmatrix}, \text{ or } \begin{pmatrix} 1 & 3 & 2 \\ 3 & 1 & 2 \end{pmatrix},$$

in fact in 3! or 6 different ways.

There are $n!$ different permutations of n objects. For the first may map onto any one of the n and, having done this, there are $(n-1)$ possibilities for the second, $(n-2)$ for the third and so on. The 6 possible permutations of three objects are

$$\begin{pmatrix} 1 & 2 & 3 \\ 1 & 2 & 3 \end{pmatrix} \begin{pmatrix} 1 & 2 & 3 \\ 1 & 3 & 2 \end{pmatrix} \begin{pmatrix} 1 & 2 & 3 \\ 2 & 3 & 1 \end{pmatrix}$$

$$\begin{pmatrix} 1 & 2 & 3 \\ 2 & 1 & 3 \end{pmatrix} \begin{pmatrix} 1 & 2 & 3 \\ 3 & 1 & 2 \end{pmatrix} \begin{pmatrix} 1 & 2 & 3 \\ 3 & 2 & 1 \end{pmatrix}.$$

Product of permutations

If we have two permutations on n objects we may define their product to be the permutation obtained by applying the first and following it by the second. Thus if x maps i into a_i, and y maps a_i into b_i, the product xy (in that order) is the permutation which maps i into b_i. This is of course a permutation, since both x and y are 1-1 mappings and therefore so is their product. Symbolically

$$x = \begin{pmatrix} 1 & 2 & \dots n \\ a_1 & a_2 & \dots a_n \end{pmatrix} \quad y = \begin{pmatrix} a_1 & a_2 & \dots a_n \\ b_1 & b_2 & \dots b_n \end{pmatrix},$$

$$xy = \begin{pmatrix} 1 & 2 & \dots n \\ a_1 & a_2 & \dots a_n \end{pmatrix} \begin{pmatrix} a_1 & a_2 & \dots a_n \\ b_1 & b_2 & \dots b_n \end{pmatrix} = \begin{pmatrix} 1 & 2 & \dots n \\ b_1 & b_2 & \dots b_n \end{pmatrix}.$$

(Note that in the above we are using the fact that the top row in the permutation y may be written in any order.)

As examples let

$$x = \begin{pmatrix} 1 & 2 & 3 \\ 3 & 1 & 2 \end{pmatrix}, \quad y = \begin{pmatrix} 1 & 2 & 3 \\ 1 & 3 & 2 \end{pmatrix}.$$

Then $$xy = \begin{pmatrix} 1 & 2 & 3 \\ 3 & 1 & 2 \end{pmatrix} \begin{pmatrix} 3 & 1 & 2 \\ 2 & 1 & 3 \end{pmatrix} = \begin{pmatrix} 1 & 2 & 3 \\ 2 & 1 & 3 \end{pmatrix}.$$

Again suppose

$$x = \begin{pmatrix} 1 & 2 & 3 & 4 \\ 2 & 3 & 4 & 1 \end{pmatrix}, \quad y = \begin{pmatrix} 1 & 2 & 3 & 4 \\ 2 & 1 & 3 & 4 \end{pmatrix}.$$

Then

$$xy = \begin{pmatrix} 1 & 2 & 3 & 4 \\ 2 & 3 & 4 & 1 \end{pmatrix} \begin{pmatrix} 2 & 3 & 4 & 1 \\ 1 & 3 & 4 & 2 \end{pmatrix} = \begin{pmatrix} 1 & 2 & 3 & 4 \\ 1 & 3 & 4 & 2 \end{pmatrix}.$$

After a little practice it is possible to omit writing down the actual product, i.e.

$$\begin{pmatrix} 1 & 2 & 3 \\ 3 & 1 & 2 \end{pmatrix} \begin{pmatrix} 3 & 1 & 2 \\ 2 & 1 & 3 \end{pmatrix}$$

in the first example. We merely note that 1 goes to 3 under x, and 3 goes to 2 under y, and so xy maps 1 onto 2. Similarly for the other objects.

Note that in the first example

$$yx = \begin{pmatrix} 1 & 2 & 3 \\ 3 & 2 & 1 \end{pmatrix}, \quad \text{i.e. } xy \neq yx.$$

It is generally true that the product is not commutative, although it may be in particular examples.

[If $$x = \begin{pmatrix} 1 & 2 & 3 \\ 2 & 3 & 1 \end{pmatrix}, \quad y = \begin{pmatrix} 1 & 2 & 3 \\ 3 & 1 & 2 \end{pmatrix},$$

it is easily seen that

$$xy = yx = \begin{pmatrix} 1 & 2 & 3 \\ 1 & 2 & 3 \end{pmatrix}.]$$

Theorem 12.6.1. *The set of all permutations of n objects under the product defined above forms a group, called the symmetric group of degree n.*

(An element of our set is a permutation; the set in question has no connection with the collection of elements which we are permuting.)

The product of two permutations is uniquely defined and is in the set of permutations.

The Associative Law is true. For suppose that x maps the object i onto a_i, y maps a_i onto b_i and z maps b_i onto c_i. Then xy maps i onto b_i and so $(xy)z$ maps i onto c_i. But x maps i onto a_i and yz maps a_i onto c_i; so $x(yz)$ also maps i onto c_i.

The identity permutation which leaves each object unchanged acts as a neutral element. For if x maps i onto a_i, being written

$$\begin{pmatrix} i \\ a_i \end{pmatrix} \text{ for short,}$$

$$ex = \begin{pmatrix} i \\ i \end{pmatrix}\begin{pmatrix} i \\ a_i \end{pmatrix} = \begin{pmatrix} i \\ a_i \end{pmatrix} = x$$

and $$xe = \begin{pmatrix} i \\ a_i \end{pmatrix}\begin{pmatrix} a_i \\ a_i \end{pmatrix} = \begin{pmatrix} i \\ a_i \end{pmatrix} = x.$$

If x is as above the permutation

$$\begin{pmatrix} a_i \\ i \end{pmatrix}$$

is its inverse.

For $$x^{-1}x = \begin{pmatrix} a_i \\ i \end{pmatrix}\begin{pmatrix} i \\ a_i \end{pmatrix} = \begin{pmatrix} a_i \\ a_i \end{pmatrix} = e$$

and $$xx^{-1} = \begin{pmatrix} i \\ a_i \end{pmatrix}\begin{pmatrix} a_i \\ i \end{pmatrix} = \begin{pmatrix} i \\ i \end{pmatrix} = e.$$

[As example, the inverse of

$$\begin{pmatrix} 1 & 2 & 3 \\ 3 & 1 & 2 \end{pmatrix} \text{ is } \begin{pmatrix} 3 & 1 & 2 \\ 1 & 2 & 3 \end{pmatrix} \text{ or } \begin{pmatrix} 1 & 2 & 3 \\ 2 & 3 & 1 \end{pmatrix}.]$$

The order of the symmetric group of degree n is $n!$, and it is non-Abelian for $n \geqslant 3$. It is often denoted by S_n.

The symmetric groups are important, particularly from a historical point of view. They were the first type of group to be studied as such, and originally 'group' meant 'group of permutations'. Many of the properties of general finite groups were discovered for the permutation groups before the abstract nature of groups was fully understood, early workers in this connection being Cayley, Cauchy and Galois, all in the nineteenth century. In a certain sense all finite groups are contained in the symmetric groups (see Cayley's theorem, §13.5).

By definition a permutation is a mapping of a set of n objects into itself which is onto and 1-1 and the set of all permutations of n objects gives all possible such mappings. We see that our definitions of product, identity and inverse for permutations agree with those given in chapter 9 for general mappings.

Symmetric groups of degree 1, 2 *and* 3

The symmetric group of degree 1 has just 1 element and is the trivial group.

The symmetric group of degree 2 has two elements,

$$e = \begin{pmatrix} 1 & 2 \\ 1 & 2 \end{pmatrix} \quad \text{and} \quad a = \begin{pmatrix} 1 & 2 \\ 2 & 1 \end{pmatrix},$$

so that $a^2 = e$. It is therefore isomorphic to the cyclic group of order 2.

The symmetric group of degree 3 has 6 elements. Let us denote

$$\begin{pmatrix} 1 & 2 & 3 \\ 2 & 3 & 1 \end{pmatrix} \quad \text{by } a$$

and

$$\begin{pmatrix} 1 & 2 & 3 \\ 1 & 3 & 2 \end{pmatrix} \quad \text{by } b.$$

We see that the other elements are given by

$$\begin{pmatrix} 1 & 2 & 3 \\ 3 & 1 & 2 \end{pmatrix} = a^2, \quad \begin{pmatrix} 1 & 2 & 3 \\ 2 & 1 & 3 \end{pmatrix} = ba, \quad \begin{pmatrix} 1 & 2 & 3 \\ 3 & 2 & 1 \end{pmatrix} = ba^2,$$

with of course

$$\begin{pmatrix} 1 & 2 & 3 \\ 1 & 2 & 3 \end{pmatrix} = e.$$

It is easy to verify that $a^3 = e$, $b^2 = e$ and that $ab = ba^2$. But these are precisely the relations for the dihedral group of order 6, and so the symmetric group of degree 3 is isomorphic to this.

We can explain this isomorphism geometrically as follows. The dihedral group is given by transformations of an equilateral triangle. Label the positions of the vertices by 1, 2, 3. (Not the vertices themselves, which move with the triangle, but the positions in space.) Corresponding to any transformation we have a permutation given by sending each of 1, 2 and 3 into the number borne by the position into which the vertex occupying the original number goes. Thus

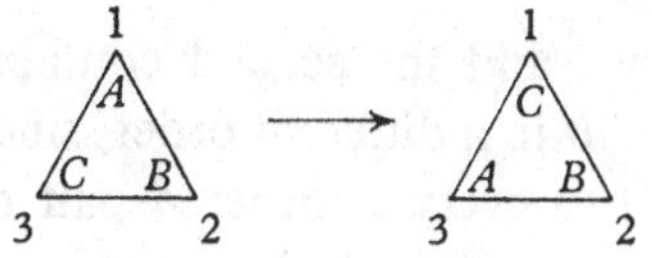

1 1
A → C corresponds to $\begin{pmatrix} 1 & 2 & 3 \\ 3 & 2 & 1 \end{pmatrix}$.
C B A B
3 2 3 2

Under this correspondence the product of two transformations corresponds to the product of their permutations, for if the first sends the vertex in position i into a_i, and the second sends that in position a_i into b_i, their product sends the vertex in position i into b_i.

It is possible to make a transformation which permutes the vertices in any of the 6 possible ways. Hence our correspondence is 1-1, and so is an isomorphism: telling us that the groups are isomorphic.

Notice that the above argument breaks down for symmetric groups of degree higher than 3. We still have a permutation corresponding to each transformation, but we can no longer obtain *all* permutations in this way, since the vertices always remain in the same order or its reversal relative to one another. Thus for a square we cannot obtain the permutation

$$\begin{pmatrix} 1 & 2 & 3 & 4 \\ 1 & 3 & 2 & 4 \end{pmatrix}.$$

Dihedral groups of order > 6 are in fact isomorphic to a portion only of the corresponding symmetric groups: to what is called a 'subgroup' of them.

Even and odd permutations

The reader may find the remainder of this section difficult to follow. The difficulty lies largely in the notation: the ideas are fairly intuitive, but it is unfortunately not possible to explain them with reasonable rigour without introducing symbolism which makes them appear more difficult than they in fact are. The work is kept as simple as possible and it is well worth while making some effort to understand it.

Suppose we have a permutation P of n objects which maps i onto a_i. Consider the products

$$\prod_{1\leqslant r<s\leqslant n}(s-r) \quad\text{and}\quad \prod_{1\leqslant r<s\leqslant n}(a_s-a_r).$$

The first product is clearly positive, and the second contains numerically every term of the first but in a different order, since both include a term corresponding to every unordered pair of the numbers $1, 2, \ldots, n$. The second product, however, will have some of its terms positive and some negative. If we denote the number of negative terms by k, the number $(-1)^k$ is called the *sign of the permutation P*, and we will denote it by $\zeta(P)$. It is equal to

$$\frac{\prod\limits_{1\leqslant r<s\leqslant n}(s-r)}{\prod\limits_{1\leqslant r<s\leqslant n}(a_s-a_r)}.$$

Considering all pairs r and s of the numbers $1, 2, \ldots, n$, we see that sometimes if $r < s$, $a_r < a_s$, while for certain pairs we have $r < s$ but $a_r > a_s$. Such an occurrence is called an *inversion*. We see that the $\zeta(P)$ is $(-1)^k$ where k is the number of inversions in P, as we range over all possible pairs (r, s). This gives us a simple practical way of calculating $\zeta(P)$.

As an example of the above, consider

$$\begin{pmatrix}1 & 2 & 3 & 4\\ 3 & 1 & 4 & 2\end{pmatrix}.$$

The inversions are given by the pairs (1, 2), (1, 4), (3, 4). Hence $k = 3$ and $\zeta(P) = (-1)^k = -1$.

If $\zeta(P) = 1$, P is called an *even permutation*, while if $\zeta(P) = -1$, P is an *odd permutation*. (These of course correspond to the number of inversions being even or odd.)

Theorem 12.6.2. *The sign of the product of two permutations is equal to the product of their signs, i.e.* $\zeta(PQ) = \zeta(P)\zeta(Q)$.

Let

$$P = \begin{pmatrix} i \\ a_i\end{pmatrix}, \quad Q = \begin{pmatrix} a_i \\ b_i\end{pmatrix}.$$

$$\zeta(P) = \frac{\prod\limits_{1\leqslant r<s\leqslant n}(s-r)}{\prod\limits_{1\leqslant r<s\leqslant n}(a_s-a_r)}, \quad \zeta(PQ) = \frac{\prod\limits_{1\leqslant r<s\leqslant n}(s-r)}{\prod\limits_{1\leqslant r<s\leqslant n}(b_s-b_r)}.$$

We now note that although $\zeta(P)$ was given as

$$\frac{\prod_{1 \leqslant r < s \leqslant n} (s-r)}{\prod_{1 \leqslant r < s \leqslant n} (a_s - a_r)}$$

above, it is also equal to

$$\frac{\prod_{1 \leqslant {r \atop s} \leqslant n,\, r \neq s} (s-r)}{\prod_{1 \leqslant {r \atop s} \leqslant n,\, r \neq s} (a_s - a_r)},$$

where either the larger or smaller of r and s is taken first, provided we still take all possible products. For if we change the sign of $(s-r)$ we also change that of $(a_s - a_r)$ and hence leave the ratio unchanged.

Hence

$$\zeta(Q) = \frac{\prod_{1 \leqslant r < s \leqslant n} (a_s - a_r)}{\prod_{1 \leqslant r < s \leqslant n} (b_s - b_r)}.$$

Thus

$$\zeta(PQ) = \frac{\prod_{1 \leqslant r < s \leqslant n} (s-r)}{\prod_{1 \leqslant r < s \leqslant n} (b_s - b_r)} = \frac{\prod_{1 \leqslant r < s \leqslant n} (s-r)}{\prod_{1 \leqslant r < s \leqslant n} (a_s - a_r)} \frac{\prod_{1 \leqslant r < s \leqslant n} (a_s - a_r)}{\prod_{1 \leqslant r < s \leqslant n} (b_s - b_r)}$$

$$= \zeta(P)\zeta(Q).$$

The identity permutation e has no inversions and hence its sign is 1. Since $P^{-1}P = e$ we have $\zeta(P^{-1}).\zeta(P) = \zeta(e) = 1$ and so $\zeta(P^{-1}) = \zeta(P)$.

By theorem 12.6.2 the product of two even or two odd permutations is even, while that of an even and an odd one is odd.

Transpositions

A transposition is a permutation which interchanges two of the objects but leaves the others unaltered. Thus

$$\begin{pmatrix} 1 & 2 & 3 & 4 \\ 1 & 4 & 3 & 2 \end{pmatrix}$$

is a transposition. A transposition always has order 2.

Theorem 12.6.3. *If P is a transposition, $\zeta(P) = -1$.*

Suppose

$$P = \begin{pmatrix} 1 & 2 \ldots r \ldots s \ldots n \\ 1 & 2 \ldots s \ldots r \ldots n \end{pmatrix}.$$

We have the following pairs giving an inversion:

$$(r, r+1), (r, r+2), \ldots, (r, s),$$
$$(r+1, s), (r+2, s), \ldots, (s-1, s).$$

Thus there are $(s-r)+(s-r-1) = (2s-2r-1)$ inversions, an odd number.

Theorem 12.6.4. *Any permutation may be expressed as the product of transpositions, but this product will not be unique.*

If

$$P = \begin{pmatrix} i \\ a_i \end{pmatrix}$$

let T_1 transpose 1 and a_1. Let T_2 transpose a_2 and the number in the second position after applying T_1, let T_3 transpose a_3 and the number in the third position after applying T_1T_2, and so on. We see that after at most $(n-1)$ transpositions we obtain P.

The product is not unique, since we may insert the pair TT, where T is any transposition. Other less trivial alterations may be made.

Example. Let
$$P = \begin{pmatrix} 1 & 2 & 3 & 4 \\ 4 & 1 & 2 & 3 \end{pmatrix}.$$

Then
$$T_1 = \begin{pmatrix} 1 & 2 & 3 & 4 \\ 4 & 2 & 3 & 1 \end{pmatrix}, \quad T_2 = \begin{pmatrix} 1 & 2 & 3 & 4 \\ 2 & 1 & 3 & 4 \end{pmatrix}$$
$$T_3 = \begin{pmatrix} 1 & 2 & 3 & 4 \\ 1 & 3 & 2 & 4 \end{pmatrix}$$

and

$$T_1T_2T_3 = \begin{pmatrix} 1 & 2 & 3 & 4 \\ 4 & 2 & 3 & 1 \end{pmatrix}\begin{pmatrix} 4 & 2 & 3 & 1 \\ 4 & 1 & 3 & 2 \end{pmatrix}\begin{pmatrix} 4 & 1 & 3 & 2 \\ 4 & 1 & 2 & 3 \end{pmatrix}.$$
$$= P.$$

Another way of expressing P as a product of transpositions is given by

$$P = \begin{pmatrix} 1 & 2 & 3 & 4 \\ 1 & 3 & 2 & 4 \end{pmatrix}\begin{pmatrix} 1 & 3 & 2 & 4 \\ 1 & 4 & 2 & 3 \end{pmatrix}\begin{pmatrix} 1 & 4 & 2 & 3 \\ 4 & 1 & 2 & 3 \end{pmatrix}.$$

Theorem 12.6.5. *In any expression of an even permutation as a product of transpositions, there are an even number of these, while for an odd permutation there are an odd number.*

The sign of the permutation is the product of the signs of the transpositions, and any transposition has sign -1.

Theorem 12.6.6. *Of all possible permutations of n objects exactly half are even and half odd* ($n \geqslant 2$).

Let A be any fixed odd permutation. (If $n \geqslant 2$ such an A exists, for example the permutation which interchanges 1 and 2 is odd.) Corresponding to any even permutation P we have an odd one AP. (Using theorem 12.6.2.) For different P we obtain different AP, since $AP_1 = AP_2 \Rightarrow P_1 = P_2$ by the Cancellation Law. Also *any* odd permutation Q is of the form AP where $P = A^{-1}Q$, since this is even as both A^{-1} and Q are odd. Hence we have a 1-1 correspondence $P \leftrightarrow AP$ between all even and all odd permutations, and so there are an equal number of each.

Theorem 12.6.7 *The* $\frac{1}{2}n!$ *even permutations of n objects form a group, called the* alternating group of degree *n*.

The product of two even permutations is even, i.e. is in the set. The identity permutation is even, as is the inverse of an even one. The Associative Law is of course satisfied, since this is so for any permutations.

Example. The alternating group of degree 3. The even permutations of three objects are easily seen to be

$$e = \begin{pmatrix} 1 & 2 & 3 \\ 1 & 2 & 3 \end{pmatrix}, \quad a = \begin{pmatrix} 1 & 2 & 3 \\ 2 & 3 & 1 \end{pmatrix}, \quad a^2 = \begin{pmatrix} 1 & 2 & 3 \\ 3 & 1 & 2 \end{pmatrix}.$$

These form a group which is identified as the cyclic group of order 3.

12.7. Transformations of solid figures

In § 12.5 we investigated the transformation groups of regular polygons: in this section we consider some solid figures. As the transformation group of a figure gives a description of its symmetry it is to be expected that the most interesting results arise from the 5 regular polyhedra. We will therefore concentrate on these, but will mention also the transformations of a sphere and another example of the cyclic and dihedral groups. Transformations of solids other than those dealt with give rise to much the same groups.

The transformation group of a regular tetrahedron (the tetrahedral group)

Theorem 12.7.1. *The tetrahedral group has* 12 *elements and is isomorphic to the alternating group of degree* 4.

To each transformation there corresponds a permutation of 4 objects as follows. Let the positions of the vertices be numbered 1, 2, 3, 4, and suppose that a certain transformation sends the vertex in position i to the position a_i. Then we take the permutation

$$\begin{pmatrix} i \\ a_i \end{pmatrix}$$

as corresponding to the transformation. Thus the transformation in the figure corresponds to

$$\begin{pmatrix} 1 & 2 & 3 & 4 \\ 2 & 1 & 4 & 3 \end{pmatrix}.$$

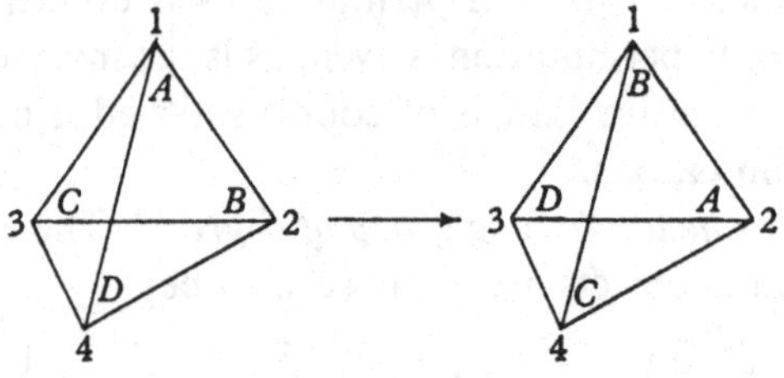

Fig. 53

Under this correspondence there is a unique permutation to each possible transformation, and each of the latter gives rise to a different permutation. The product of permutations corresponds to the product of the allied transformations and of course the neutral elements and the inverses also correspond. There is not, however, a transformation corresponding to every possible permutation. Let us investigate which transformations exist.

If we imagine the vertex 1 fixed we have 3 possible transformations (see figure 54). We notice that these are all even permutations. The 3 odd ones which leave 1 unchanged cannot be obtained by moving the tetrahedron, since to bring the vertices B, C, D into the requisite positions would involve A

being moved to underneath the triangle BCD; thus the tetrahedron would no longer occupy the same place.

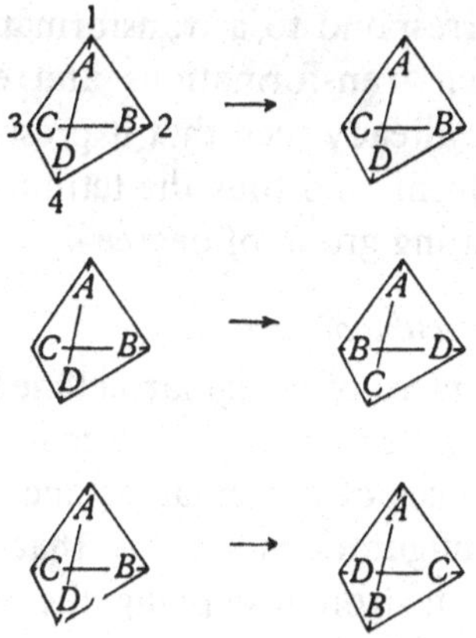

corresponding to $\begin{pmatrix} 1 & 2 & 3 & 4 \\ 1 & 2 & 3 & 4 \end{pmatrix}$

corresponding to $\begin{pmatrix} 1 & 2 & 3 & 4 \\ 1 & 3 & 4 & 2 \end{pmatrix}$

corresponding to $\begin{pmatrix} 1 & 2 & 3 & 4 \\ 1 & 4 & 2 & 3 \end{pmatrix}$

Fig. 54

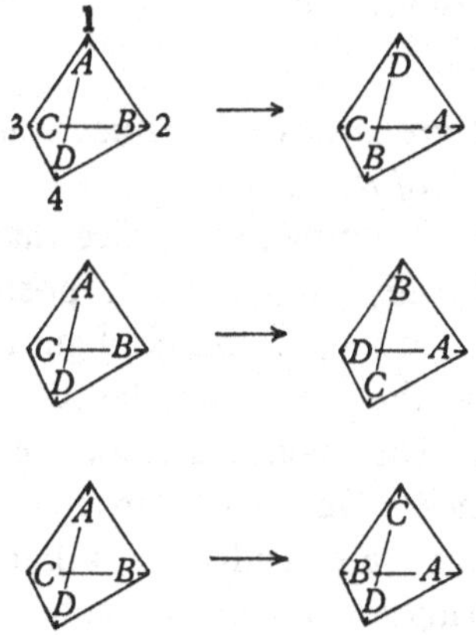

corresponding to $\begin{pmatrix} 1 & 2 & 3 & 4 \\ 2 & 4 & 3 & 1 \end{pmatrix}$

corresponding to $\begin{pmatrix} 1 & 2 & 3 & 4 \\ 2 & 1 & 4 & 3 \end{pmatrix}$

corresponding to $\begin{pmatrix} 1 & 2 & 3 & 4 \\ 2 & 3 & 1 & 4 \end{pmatrix}$

Fig. 55

Now the vertex at 1 may be moved to any of the other three positions, and corresponding to each there are exactly 3 possible configurations of the other 3 vertices, obtained as in the first case above by rotations. Thus there are 12 elements in the transformation group. We give in figure 55 the 3 cases where the vertex at 1 is moved to 2, together with the corresponding permutations.

We see again that all these permutations are even. It can readily be verified that the remaining 6 cases also correspond to even permutations. But there are only 12 even permutations of 4 objects, and hence all these correspond to a transformation. Thus the correspondence between transformations and even permutations is 1-1, and we have already seen that it preserves product. Hence it is an isomorphism, and thus the tetrahedral group is isomorphic to the alternating group of degree 4.

The octahedral and hexahedral (*or cubical*) *groups*

If we take the 8 centres of the faces of a regular octahedron and join them we obtain a cube, and conversely the 6 centres of the faces of a cube form the vertices of a regular octahedron. Thus taking either of these configurations we see that any movement which leaves the octahedron occupying the same position has the same effect on the cube, and vice versa. Hence the two groups are the same. It is usually called the *octahedral group*, although we will find it convenient to consider it as the transformation group of a cube.

Theorem 12.7.2. *The octahedral group has* 24 *elements and is isomorphic to the symmetric group of degree* 4.

Considering the group as that of a cube, we notice that any vertex may be transformed into any of 8 possible positions. Given one of these positions there are 3 possible configurations, since the 3 squares which meet at the vertex may be rotated to give a transformation in 3 ways. The diagrams show the cases where the vertex A is unchanged. Hence the group has 24 elements. To examine its structure we consider the 4 diagonals of the cube. Since the transformation leaves the cube in the same position in space the diagonals are permuted among themselves. Each transformation therefore corresponds to a permutation of 4 objects, and we will show below that different transformations give rise to different permutations. Hence the 24 possible transformations correspond to 24 *different* permutations, i.e. to the totality of possible ones. But this correspondence clearly preserves the product structure and so is an isomorphism, being 1-1. Hence the group is isomorphic to the symmetric group of degree 4.

It remains to show that different transformations cannot correspond to the same permutation of diagonals. We will show first that the only one corresponding to the identity permutation, i.e. leaving all 4 diagonals unaltered in position, is the identity transformation. For if a diagonal is left unaltered it must either have its ends unchanged or interchanged. In the first case the transformation must be a rotation about the diagonal in question, while in the second it must be about some

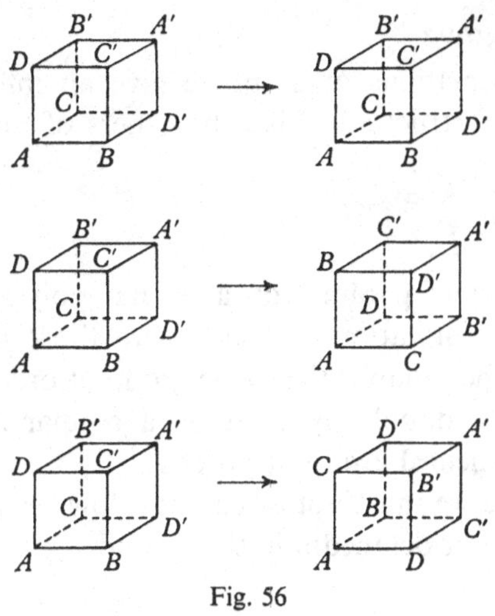

Fig. 56

line perpendicular to the diagonal. Thus if all 4 diagonals are left in the same positions the transformation must be a rotation perpendicular to at least 3 of them, since it cannot be *about* more than 1. But there is no line perpendicular to 3 diagonals, and so the transformation is the identical one. Now suppose that the transformations a and b both correspond to the same permutation P. Then ab^{-1} corresponds to $PP^{-1} = I$, and so by what we have proved above $ab^{-1} = e$, the identical transformation. Thus $a = b$ and we have proved the theorem.

The dodecahedral and icosahedral groups

As for the octahedron and cube we notice that the dodecahedron and icosahedron are dual (the centres of the faces of one are the vertices of the other), and so their groups are the same.

(Note that the tetrahedron is self-dual.) The group has 60 elements, for a vertex of the dodecahedron may be transformed into any of 20 positions, and to each of these there correspond 3 possible configurations, as there are 3 pentagons meeting at the vertex. This group is in fact isomorphic to the alternating group of degree 5, but we will not prove this here. (A proof is given in *Introduction to the Theory of Finite Groups*, by W. Ledermann.)

The spherical group

The transformations of a sphere give an infinite group of highly complex structure. Like the others of this section it is non-Abelian.

Groups of pyramids

Consider a right pyramid on a regular polygonal base of n sides. Its transformations coincide with those of a one-sided n-gon and so the group is the cyclic group of order n.

If we have a double pyramid on a regular n-gon base its group is the dihedral group of order $2n$.

In the above we must not of course allow regular tetrahedra in the first case or octahedra in the second.

12.8. Some other groups

Cross-ratios

If one of the 6 different cross-ratios of 4 numbers is denoted by x, the other 5 are given by the functions

$$A(x) = 1-x, \quad B(x) = 1/x, \quad C(x) = 1/(1-x),$$

$$D(x) = (x-1)/x, \quad E(x) = x/(x-1).$$

We may define a product of any two of these (including the function x itself which we will denote by $I(x)$) as the function obtained by performing them in order: thus $(PQ)(x) = Q(P(x))$: this will always be in the set since our original cross-ratio x could have been chosen to be any of the 6, i.e. could have been $P(x)$, and so the function Q of this must be a cross-ratio.

The product is associative since

$$((PQ)R)(x) = (P(QR))(x)$$
$$= R(Q(P(x))).$$

$I(x)$ is a neutral element.

Each function has an inverse, as can be seen from the fact that, operating on $P(x)$ by each in turn we must obtain all 6 of the cross-ratios once each, and so one of them gives $I(x)$, say $PQ = I$; similarly there is a function R such that $RP = I$, whence $RPQ = Q$ and $= R$, so this common element is the inverse of P.

Hence the 6 cross-ratios form a group.

Theorem 12.8.1. *The group of cross-ratios is isomorphic to the symmetric group of degree* 3, *i.e. to the dihedral group of order* 6.

$$C^2(x) = 1\bigg/\left(1-\frac{1}{1-x}\right) = \frac{1-x}{-x} = \frac{x-1}{x} = D(x),$$

$$C^3(x) = (DC)(x) = C(D(x)) = 1\bigg/\left(1-\frac{x-1}{x}\right) = \frac{x}{1} = I(x),$$

$$B^2(x) = 1\bigg/\left(\frac{1}{x}\right) = I(x),$$

$$(BC)(x) = C(B(x)) = 1\bigg/\left(1-\frac{1}{x}\right) = \frac{x}{x-1} = E(x),$$

$$(BC^2)(x) = (BD)(x) = D(B(x)) = \left(\frac{1}{x}-1\right)\bigg/\left(\frac{1}{x}\right) = 1-x$$
$$= A(x).$$

Hence the elements are I, C, C^2, B, BC and BC^2, with $C^3 = B^2 = I$. Furthermore

$$(CB)(x) = B(C(x)) = 1\bigg/\left(\frac{1}{1-x}\right) = A(x)$$

and so $CB = BC^2$.

The result follows, since we have the relations which define the symmetric group.

A group from set theory

If we take the collection of all possible subsets of some given set S, each subset being considered as an element of our collection, we might expect that under suitable definitions of product our collection would form a group.

The natural definitions to try are $\cup$ and $\cap$. These are both associative, the elements $\emptyset$ and S act as respective neutral elements, but it is not possible to have inverses, since there is no element B such that $A \cup B = \emptyset$ or $A \cap B = S$. Thus we do not have a group under either $\cup$ or $\cap$.

Let us define the product AB as $(A \cup B)-(A \cap B)$, as shown in the Venn diagram of figure 57.

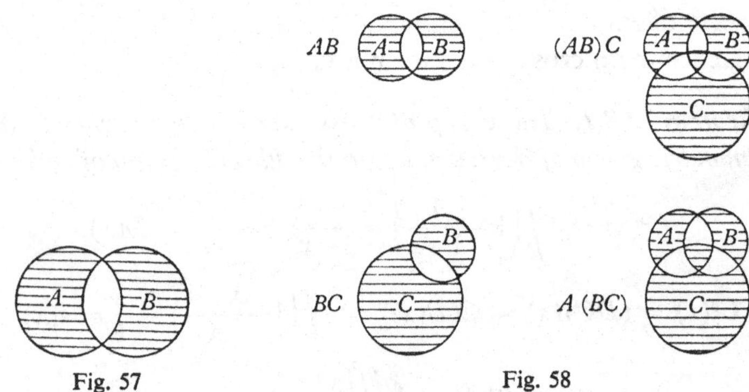

Fig. 57

Fig. 58

This product is associative as illustrated, the product ABC comprising those elements of S which are in either 1 or all 3 of the subsets A, B and C.

The set $\emptyset$ is a neutral element, since

$$\begin{aligned} A\emptyset = \emptyset A &= (A \cup \emptyset)-(A \cap \emptyset) \\ &= A-\emptyset \\ &= A. \end{aligned}$$

The inverse of A is A itself, since

$$\begin{aligned} AA &= (A \cup A)-(A \cap A) \\ &= A-A \\ &= \emptyset. \end{aligned}$$

Hence under this product our collection forms a group; this is Abelian and every element except e $(= \emptyset)$ has order 2.

The group of a polynomial

Consider the polynomial $x_1x_2+x_3+x_4$ in the four variables x_1, x_2, x_3, x_4. If we permute the x_i's in certain ways the polynomial is left unaltered: we could, for example, interchange x_1 and x_2, leaving x_3 and x_4 unaltered, or we could interchange both pairs, but we may not interchange x_1 and x_3. Considering the set of all permutations which leave the polynomial invariant, we see that the product of two such is in the set, the identity is in it, as is the inverse of any permutation in the set. Thus the subset of the symmetric group of degree 4 which leaves this polynomial invariant forms a group, known as the *group of the* polynomial.

In the example given it is easily seen that the group consists of the following permutations

$$e = \begin{pmatrix} 1 & 2 & 3 & 4 \\ 1 & 2 & 3 & 4 \end{pmatrix}, \quad a = \begin{pmatrix} 1 & 2 & 3 & 4 \\ 2 & 1 & 3 & 4 \end{pmatrix},$$

$$b = \begin{pmatrix} 1 & 2 & 3 & 4 \\ 1 & 2 & 4 & 3 \end{pmatrix}, \quad c = \begin{pmatrix} 1 & 2 & 3 & 4 \\ 2 & 1 & 4 & 3 \end{pmatrix}.$$

Since $a^2 = b^2 = e$ and $ab = c = ba$ we see that the group is the Vierergruppe.

The above may be generalised. Let $f(x_1, x_2, \ldots, x_n)$ be a polynomial in the n variables $x_1, x_2, \ldots, x_n$. Then the set of all permutations

$$\begin{pmatrix} i \\ a_i \end{pmatrix}$$

which leave f invariant, i.e. such that

$$f(x_{a_1}, x_{a_2}, \ldots, x_{a_n}) = f(x_1, x_2, \ldots, x_n),$$

form a group, the group of the polynomial f. Such a group gives a measure and description of the symmetry of the polynomial: if the group of f is the whole symmetric group of degree n we say that f is *symmetrical* in $x_1, x_2, \ldots, x_n$.

An artificial group

As an example of a group constructed completely artificially we give the following.

An element is the ordered pair (a, b) where a and b are real numbers with a non-zero. The product is defined as

$$(a, b).(c, d) = (ac, bc+d).$$

Since $a \neq 0$ and $c \neq 0 \Rightarrow ac \neq 0$ the product is in the set.

Product is associative, since

$$(a, b)((c, d)(e, f)) = (a, b)(ce, de+f) = (ace, bce+de+f),$$

$$((a, b)(c, d))(e, f) = (ac, bc+d)(e, f) = (ace, bce+de+f).$$

$(1, 0)$ is a neutral element, since

$$(1, 0)(a, b) = (a, 0a+b) = (a, b),$$

$$(a, b)(1, 0) = (a, b1+0) = (a, b).$$

The inverse of (a, b) is $(a^{-1}, -ba^{-1})$ as can be easily verified and, since a is not zero, this exists and is in the set since a^{-1} is not zero.

Thus the elements form a group under this product. The group is not Abelian since $(c, d)(a, b) = (ac, ad+b)$, and this is not the same as $(a, b)(c, d)$.

12.9. Direct products

The group of two-dimensional vectors is formed from that of real numbers (both under addition) in the following way. We take a vector to be an ordered pair of real numbers (x, y) and form the sum of two vectors by adding the corresponding co-ordinates. If we consider the real numbers as the set of points on a straight line we obtain a representation of the set of vectors by drawing two lines at right angles, the axes, and then representing a vector by a point in their plane. The group of vectors may be thought of as a 'product' of two groups of reals.

The same idea may be extended to any pair of groups. Suppose we have any two groups G and H and consider the set of all ordered pairs (x, y) where $x \in G$ and $y \in H$. Define a product of two pairs by $(x_1, y_1).(x_2, y_2) = (x_1x_2, y_1y_2)$, the products x_1x_2 and y_1y_2 being those defined in G and H, respectively. Then under this product the set of ordered pairs forms a group. For the product is unique and in the set. It is associative since the products defined in G and H are: we see that

$$(x_1, y_1)(x_2, y_2)(x_3, y_3) = (x_1x_2x_3, y_1y_2y_3).$$

If e and f are the respective neutral elements of G and H it is easily seen that (e, f) acts as a neutral element in our set of pairs, while the inverse of (x, y) is given by (x^{-1}, y^{-1}).

The group we have defined is known as the *direct product* of G and H and is denoted by $G \times H$. It gives us a powerful method of forming new groups from known ones and, in more advanced work, of investigating the structure of a given group. We see that if R denotes the group of reals, the group of two-dimensional vectors is (strictly speaking is isomorphic to) $R \times R$.

G and H may be any groups, one or both finite, one or both Abelian. If they are both finite of orders m and n, the order of $G \times H$ is mn. $G \times H$ is Abelian if and only if both G and H are.

We may extend the idea to 3 or more groups, in a fairly obvious manner. The direct product of the n groups $G_1, G_2, \ldots, G_n$, written $G_1 \times G_2 \times G_3 \times \ldots \times G_n$, is the set of all ordered n-tuples $(g_1, g_2, \ldots, g_n)$, $g_i \in G_i$, with product defined as

$$(g_1, g_2, \ldots, g_n) . (g_1', g_2', \ldots, g_n') = (g_1g_1', g_2g_2', \ldots, g_ng_n').$$

It is left to the reader to verify the group axioms for this set.

Note that $G \times H$ and $H \times G$ are isomorphic, and

$$(G \times H) \times K \cong G \times (H \times K) \cong G \times H \times K.$$

We see that the group of n-dimensional vectors is isomorphic to $R \times R \times \ldots \times R$, containing n terms, as is the group of polynomials of degree $\leqslant n-1$ under addition, since this latter is isomorphic to the group of n-dimensional vectors.

Example 1. The Vierergruppe is isomorphic to $C_2 \times C_2$, where C_2 is the cyclic group of order 2.

If the elements of C_2 are f and x let

$$e = (f, f),$$
$$a = (f, x),$$
$$b = (x, f).$$

Then $ab = (x, x)$ and $a^2 = b^2 = e, ab = ba$. Thus $C_2 \times C_2 \cong$ the Vierergruppe.

Example 2. If C_n denotes the cyclic group of order n, $C_2 \times C_3 \cong C_6$.

If the elements of C_2 are e and x, and those of C_3 are f, y, y^2 so that $x^2 = e, y^3 = f$, then if $a = (x, y)$ it is easy to see that the other elements of $C_2 \times C_3$ are given by the neutral element i and by a^2, a^3, a^4, a^5 with $a^6 = i$.

12.10. Generators and relations

In describing the symmetric group of degree 3 we have several times used the fact that all its elements may be expressed in terms of two of them, a and b, and that these satisfy the equations $a^3 = b^2 = e$ and $ab = ba^2$.

A set of elements of a group which have the property that *all* the elements may be expressed in terms of them or their inverses is called a *set of generators* for the group. If none of them can be expressed in terms of the others they are *independent* generators, and in this case we cannot omit any of them and still have a set of generators.

Any power, positive or negative, of a generator a can be expressed in terms of a or a^{-1}, as can $e = a^0 = aa^{-1}$. If $\{a_1, a_2, ..., a_n\}$ are a set of generators of an Abelian group any element of the group may be expressed in the form $a_1^{\lambda_1} a_2^{\lambda_2} \ldots a_n^{\lambda_n}$, but in the non-Abelian case we may have more complicated expressions such as, for instance, $a_1^2 a_2 a_1 a_2^{-1}$.

Any finite group possesses a set of generators, viz. the set of all elements of the group, and this may be reduced to an independent set by omitting any that are expressible in terms of the rest. The independent set need not be unique: in the symmetric group of degree 3 both $\{a, b\}$ and $\{a, ba\}$ have the property, and there are other pairs.

An infinite group may or may not possess a finite set of generators. Thus the group of integers is generated by the element 1, while the group of Gaussian integers has the pair of generators $\{1, i\}$. On the other hand there is no finite set of elements which generate the group of real numbers under addition, or the group of rationals under addition.

Relations such as $a^3 = e$, $b^2 = e$, $ab = ba^2$ which exhibit the structure in terms of generators are called *defining relations*. The 3 given are sufficient to describe completely the symmetric group of degree 3, and are all independent.

A useful method of defining a group is by giving an independent set of generators, together with their corresponding defining relations. This has already been done for the symmetric group of degree 3, the dihedral groups and the Vierergruppe. For the dihedral group of order $2n$ we have 2 generators a and b and the relations $a^n = e$, $b^2 = e$, $ab = ba^{n-1}$. For the Vierergruppe these become $a^2 = b^2 = e$ and $ab = ba$. (The last may be replaced by $(ab)^2 = e$, showing that the defining relations need not be unique.) The cyclic group of order n is generated by x with $x^n = e$, while the infinite cyclic group is generated by x with no defining relation.

We may define a group by laying down an arbitrary set of generators and relations. There is no guarantee, however, that these are independent, and we may find that simpler relations can be deduced, sometimes making the group very simple indeed. As an example of this we refer to worked exercise 4 on p. 215. Here we laid down seemingly complicated and interesting relations between 2 elements a and b, which we will now consider as generators. We could expect to derive an interesting group, but in fact we saw that $a^2 = b^3 = e$ and $ab = ba$. Hence our group has just 6 elements e, a, b, ab, b^2, ab^2 and, by considering the powers of ab, we see that it is merely the cyclic group order 6 generated by ab.

Worked exercises

1. Identify the group of residues modulo 16 prime to 16 under multiplication.

The residues prime to 16 are 1, 3, 5, 7, 9, 11, 13, 15. Let $a = 3$. Then $a^2 = 9$, $a^3 = 11$, $a^4 = 1$. Let $b = 7$. Then $b^2 = 1$, $ab = 5$, $a^2b = 15$, $a^3b = 13$. Hence the 8 elements are e, a, a^2, a^3, b, ab, a^2b, a^3b with $a^4 = b^2 = e$ and of course $ab = ba$ since the group is Abelian.

Thus the group is isomorphic to $C_4 \times C_2$ where C_4 is cyclic of order 4 with generator α say, and C_2 is cyclic of order 2 with generator β. The isomorphism is given by the correspondence $a \leftrightarrow (\alpha, f)$, $b \leftrightarrow (e, \beta)$.

(Note that in the above we did not take $b = 5$, the smallest element which is not a power of a. It is true that the 4 remaining elements could still be expressed in the form b, ab, a^2b, a^3b but we no longer have that $b^2 = 1$ and so can no longer obtain an isomorphism easily. We could have chosen 15 for b, but not 13.)

2. Give the orders and inverses of the dihedral group of order 10.

Since $a^5 = e$ the orders and inverses of the powers of a are:

$$\left.\begin{matrix} e \\ a \\ a^2 \\ a^3 \\ a^4 \end{matrix}\right\} \text{ has order } \left\{\begin{matrix} 1 \\ 5 \\ 5 \\ 5 \\ 5 \end{matrix}\right\} \text{ and inverse } \left\{\begin{matrix} e, \\ a^4, \\ a^3, \\ a^2, \\ a. \end{matrix}\right.$$

Now consider ba^r, $0 \leqslant r \leqslant 4$. By theorem 12.5.1, $(ba^r)^2 = b.a^rb.a^r = b.ba^{5-r}.a^r = b^2a^5 = e$. Hence all 5 elements have order 2 and thus are their own inverses.

3. If

$$a = \begin{pmatrix} 1 & 2 & 3 & 4 & 5 \\ 5 & 3 & 4 & 2 & 1 \end{pmatrix}, \quad b = \begin{pmatrix} 1 & 2 & 3 & 4 & 5 \\ 4 & 1 & 3 & 5 & 2 \end{pmatrix}, \quad c = \begin{pmatrix} 1 & 2 & 3 & 4 & 5 \\ 1 & 3 & 2 & 5 & 4 \end{pmatrix}$$

find abc and ca^2 and determine whether $a^{-1}c$ is even or odd.

$$abc = \begin{pmatrix} 1 & 2 & 3 & 4 & 5 \\ 5 & 3 & 4 & 2 & 1 \end{pmatrix}\begin{pmatrix} 5 & 3 & 4 & 2 & 1 \\ 2 & 3 & 5 & 1 & 4 \end{pmatrix}\begin{pmatrix} 2 & 3 & 5 & 1 & 4 \\ 3 & 2 & 4 & 1 & 5 \end{pmatrix}$$

$$= \begin{pmatrix} 1 & 2 & 3 & 4 & 5 \\ 3 & 2 & 4 & 1 & 5 \end{pmatrix},$$

$$a^2 = \begin{pmatrix} 1 & 2 & 3 & 4 & 5 \\ 5 & 3 & 4 & 2 & 1 \end{pmatrix}\begin{pmatrix} 5 & 3 & 4 & 2 & 1 \\ 1 & 4 & 2 & 3 & 5 \end{pmatrix} = \begin{pmatrix} 1 & 2 & 3 & 4 & 5 \\ 1 & 4 & 2 & 3 & 5 \end{pmatrix}.$$

Hence

$$ca^2 = \begin{pmatrix} 1 & 2 & 3 & 4 & 5 \\ 1 & 3 & 2 & 5 & 4 \end{pmatrix}\begin{pmatrix} 1 & 3 & 2 & 5 & 4 \\ 1 & 2 & 4 & 5 & 3 \end{pmatrix} = \begin{pmatrix} 1 & 2 & 3 & 4 & 5 \\ 1 & 2 & 4 & 5 & 3 \end{pmatrix}.$$

(In the above working the middle step may be left out when the process is familiar and the answer written down immediately by considering what happens to 1, 2, 3, ... in turn.)

The inversions of a are (12), (13), (14), (15), (24), (25), (34), (35), (45), i.e. 9 in number. Hence a is odd. c is the product of the two transpositions

$$\begin{pmatrix} 1 & 2 & 3 & 4 & 5 \\ 1 & 3 & 2 & 4 & 5 \end{pmatrix} \text{ and } \begin{pmatrix} 1 & 2 & 3 & 4 & 5 \\ 1 & 2 & 3 & 5 & 4 \end{pmatrix}.$$

and so is even. Since a is odd, so is a^{-1} and hence $a^{-1}c$ is odd also.

4. Find the group of the polynomial $x_1x_3+x_1x_4+x_2x_3+x_2x_4$.

The polynomial is $(x_1+x_2)(x_3+x_4)$. It is unchanged if we transpose x_1 and x_2, or x_3 and x_4, or if we transpose *both* x_1 and x_3, and x_2 and x_4. Any product of these three permutations leaves the polynomial invariant, and these are the only independent permutations which do so, and so the group is generated by

$$a = \begin{pmatrix} 1 & 2 & 3 & 4 \\ 2 & 1 & 3 & 4 \end{pmatrix}, \quad b = \begin{pmatrix} 1 & 2 & 3 & 4 \\ 1 & 2 & 4 & 3 \end{pmatrix}, \quad c = \begin{pmatrix} 1 & 2 & 3 & 4 \\ 3 & 4 & 1 & 2 \end{pmatrix}.$$

We easily see that $a^2 = b^2 = c^2 = e$, and $ab = ba$. We also see that

$$ac = cb = \begin{pmatrix} 1 & 2 & 3 & 4 \\ 4 & 3 & 1 & 2 \end{pmatrix}, \quad bc = ca = \begin{pmatrix} 1 & 2 & 3 & 4 \\ 3 & 4 & 2 & 1 \end{pmatrix}.$$

Thus in any product of a, b and c we can always take c to the right-hand position and, since $ab = ba$, a can be taken to the left. Since a, b and c all have order 2 we see that the group has 8 elements

$$e, \quad a, \quad b, \quad ab, \quad c, \quad ac, \quad bc, \quad abc.$$

We now search for other relations. The group is non-Abelian and so cannot be the direct sum of cyclic groups. We know that the dihedral group of order 8 is non-Abelian and so we bear this in mind as a possibility and look for an element of order 4. (In point of fact there are only 2 non-Abelian groups of order 8 and the second of these, the so-called *quaternion group* has only one element of order 2. Thus our group *must* be the dihedral group. The reader is not expected to know these facts but, since we have at least 4 elements of order 2, viz. a, b, c, ab, and the dihedral group has 5 such elements, viz. all those which involve a turning over plus 'a^2', it seems likely that the two are isomorphic and further investigation soon settles the matter.)

We search then for an element of order 4. Consider ac.

$$(ac)^2 = acac = abc^2 = ab,$$
$$(ac)^3 = abac = a^2bc = bc,$$
$$(ac)^4 = bcac = bbcc = e.$$

Thus ac has order 4. Write $x = ac$. Then $x^2 = ab$, $x^3 = bc$. Now $ax = a^2c = c$, $ax^2 = a^2b = b$, $ax^3 = abc$. Also $xa = aca = abc = ax^3$. Hence we have our 8 elements expressed in terms of x and a with the relations of the dihedral group of order 8, and so our group is isomorphic to this.

Exercises 12A

In these exercises and in **12B** we write C_n for the cyclic group of order n, D_{2n} for the dihedral group of order $2n$, and S_n for the symmetric group of degree n.

Which of the structures **1–10** are groups? If they are not state which axioms are not satisfied; if they are groups prove the axioms and identify the groups where possible.

1. The irrational numbers under addition.

2. All rational numbers with odd denominators, when in their lowest terms, under addition, including the integers.

3. All complex numbers $a+bi$ with $b \leqslant 0$ under addition.

4. The n roots of $x^n = 1$ under multiplication.

5. The residues 0, 2, 4, 6 modulo 8 under addition.

6. The residues 0, 2, 4, 6 modulo 7 under addition.

7. The residues 1, 3, 4, 5, 9 modulo 11 under multiplication.

8. All polynomials of the form ax^3+bx+c, a, b, c real, under addition.

9. All transformations of a regular icosahedron which leave one vertex fixed.

10. If 1234 is a face of a cube, all transformations of the cube which send 1 into one of the positions 1, 2, 3 or 4.

11. Prove that in the group of rationals except 0 under multiplication the only elements of finite order are 1 and -1.

12. One isomorphism between the group $\{\pm 1, \pm i\}$ and C_4 is given by the correspondence $(1, i, -1, -i) \leftrightarrow (0, 1, 2, 3)$. Give another isomorphism.

13. Give an isomorphism between the positive real numbers under multiplication and all reals under addition.

14. Which elements of the group of complex numbers of unit modulus (under multiplication) have order n?

15. Give all the generators, and the orders of all other elements, of the groups of residues, under addition, modulo: (i) 5, (ii) 8, (iii) 12.

16. Prove that, if there exists an element order n in a group of order n, then the group is cyclic.

17. Give the multiplication tables for the following groups of residues (except 0) under multiplication and identify the groups: (i) modulo 7, (ii) modulo 11.

18. Show that in the group of residues modulo 13 under multiplication the residue 2 has order 12, and hence show that the group $\cong C_{12}$. Which other elements have order 12?

In **19–23** we denote the group of residues modulo n prime to n under multiplication by R_n.

19. In R_n prove that the residue $n-r$ is in the group if r is. Show also that $n-1$ is always in the group and has order 2.

20. Make a multiplication table for R_8 and find the orders and inverses of all its elements. Show that it is isomorphic to the Vierergruppe.

21. Repeat exercise 20 for R_{10} and show that $R_{10} \cong C_4$. What are the possible generators?

22. Repeat exercise 20 for R_{12} and show that $R_{12} \cong$ Vierergruppe.

23. Repeat exercise 20 for R_{14} and show that $R_{14} \cong C_6$.

24. In D_{10} let the elements a and b be as shown in figure 59, and let O be the centre of the pentagon.

Express the elements which turn the pentagon over about OB, OC, OD, OE in terms of a and b and verify by using the defining relations of D_{10} that each of these elements has order 2.

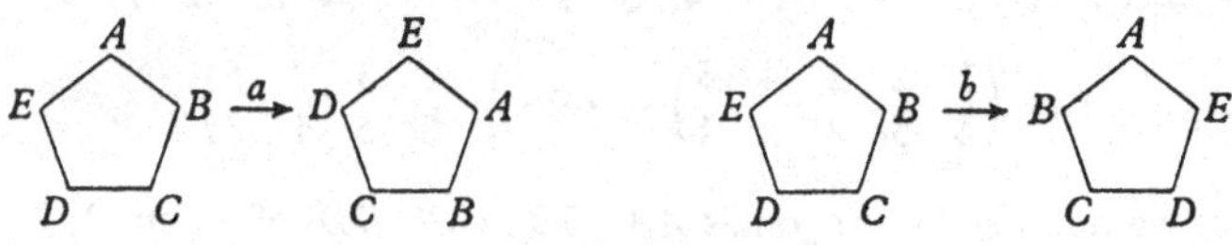

Fig. 59

25. For D_{12} prove the following from the defining relations and illustrate by figures showing the transformations:

$$\text{(i) } (ba^2)^2 = e, \quad \text{(ii) } aba = b, \quad \text{(iii) } ba^3b = a^3.$$

26. Give the orders and inverses of all elements of D_{12}.

27. Make a multiplication table for D_8.

28. Find the orders of the following elements of S_4 and state whether they are even or odd:

$$\text{(i) } \begin{pmatrix} 1 & 2 & 3 & 4 \\ 2 & 1 & 4 & 3 \end{pmatrix}, \quad \text{(ii) } \begin{pmatrix} 1 & 2 & 3 & 4 \\ 3 & 1 & 4 & 2 \end{pmatrix}, \quad \text{(iii) } \begin{pmatrix} 1 & 2 & 3 & 4 \\ 4 & 2 & 3 & 1 \end{pmatrix}.$$

29. Find the orders of the following elements of S_5 and state whether they are even or odd:

$$\text{(i) } \begin{pmatrix} 1 & 2 & 3 & 4 & 5 \\ 5 & 4 & 3 & 2 & 1 \end{pmatrix}, \quad \text{(ii) } \begin{pmatrix} 1 & 2 & 3 & 4 & 5 \\ 5 & 1 & 2 & 4 & 3 \end{pmatrix}, \quad \text{(iii) } \begin{pmatrix} 1 & 2 & 3 & 4 & 5 \\ 3 & 5 & 4 & 2 & 1 \end{pmatrix}.$$

30. If
$$a = \begin{pmatrix} 1 & 2 & 3 & 4 \\ 2 & 3 & 4 & 1 \end{pmatrix}, \quad b = \begin{pmatrix} 1 & 2 & 3 & 4 \\ 1 & 3 & 2 & 4 \end{pmatrix}.$$

find (i) ab, (ii) ba, (iii) a^2b, (iv) $(ab)^2$, (v) aba^3.

31. If
$$a = \begin{pmatrix} 1 & 2 & 3 & 4 \\ 2 & 1 & 4 & 3 \end{pmatrix}, \quad b = \begin{pmatrix} 1 & 2 & 3 & 4 \\ 4 & 3 & 2 & 1 \end{pmatrix}, \quad c = \begin{pmatrix} 1 & 2 & 3 & 4 \\ 3 & 1 & 2 & 4 \end{pmatrix}$$

show that $cac^{-1} = b$.

32. If
$$a = \begin{pmatrix} 1 & 2 & 3 & 4 & 5 \\ 3 & 5 & 2 & 4 & 1 \end{pmatrix}, \quad b = \begin{pmatrix} 1 & 2 & 3 & 4 & 5 \\ 3 & 2 & 1 & 5 & 4 \end{pmatrix}, \quad c = \begin{pmatrix} 1 & 2 & 3 & 4 & 5 \\ 1 & 3 & 2 & 5 & 4 \end{pmatrix}.$$

find (i) ab, (ii) ac, (iii) abc, (iv) cba, (v) $abcb$.

33. Express the following as products of transpositions and hence determine whether they are even or odd:

$$\text{(i) } \begin{pmatrix} 1 & 2 & 3 \\ 3 & 2 & 1 \end{pmatrix}, \quad \text{(ii) } \begin{pmatrix} 1 & 2 & 3 & 4 \\ 2 & 1 & 4 & 3 \end{pmatrix}, \quad \text{(iii) } \begin{pmatrix} 1 & 2 & 3 & 4 & 5 \\ 3 & 4 & 2 & 1 & 5 \end{pmatrix},$$

$$\text{(iv) } \begin{pmatrix} 1 & 2 & 3 & 4 & 5 \\ 2 & 4 & 5 & 1 & 3 \end{pmatrix}, \quad \text{(v) } \begin{pmatrix} 1 & 2 & 3 & 4 & 5 & 6 \\ 4 & 6 & 3 & 5 & 1 & 2 \end{pmatrix}.$$

34. Prove that an odd permutation must have even order.

35. Determine the number of inversions in the following and hence find whether they are even or odd:

$$\text{(i)} \begin{pmatrix} 1 & 2 & 3 \\ 3 & 1 & 2 \end{pmatrix}, \quad \text{(ii)} \begin{pmatrix} 1 & 2 & 3 & 4 \\ 1 & 4 & 3 & 2 \end{pmatrix}, \quad \text{(iii)} \begin{pmatrix} 1 & 2 & 3 & 4 \\ 4 & 2 & 3 & 1 \end{pmatrix},$$

$$\text{(iv)} \begin{pmatrix} 1 & 2 & 3 & 4 & 5 \\ 4 & 5 & 3 & 2 & 1 \end{pmatrix} \quad \text{(v)} \begin{pmatrix} 1 & 2 & 3 & 4 & 5 \\ 4 & 3 & 5 & 1 & 2 \end{pmatrix}.$$

36. If we number the diagonals AA', BB', CC', DD' of a cube 1, 2, 3, 4 respectively draw diagrams to illustrate the transformations corresponding to the permutations of the diagonals given by:

$$\text{(i)} \begin{pmatrix} 1 & 2 & 3 & 4 \\ 2 & 1 & 4 & 3 \end{pmatrix}, \quad \text{(ii)} \begin{pmatrix} 1 & 2 & 3 & 4 \\ 3 & 2 & 4 & 1 \end{pmatrix}, \quad \text{(iii)} \begin{pmatrix} 1 & 2 & 3 & 4 \\ 2 & 3 & 4 & 1 \end{pmatrix}.$$

37. What are the transformation groups of the following: (i) a rhombus; (ii) a rectangular box; (iii) a prism whose cross-section is a regular pentagon?

38. Considering the group of §12.8 let the set S consist of 2 members α and β. How many subsets has this, including S and $\varnothing$? Give a multiplication table for the group under $AB = (A \cup B) - (A \cap B)$ and identify it.

39. Repeat exercise 38 for the set S consisting of 3 members α, β, γ.

In **40–45** list the permutations which leave the polynomials invariant, make a multiplication table for them and identify the group of the polynomials.

40. $x_1x_2 + x_1x_3$.

41. $x_1x_2 + x_3x_4$.

42. $2x_1x_2x_3 + x_4$.

43. $x_1x_2 + x_3x_4x_5$.

44. $x_1x_2x_5 + x_3x_4x_5$.

45. Prove that the group of Gaussian integers $\cong Z \times Z$ where Z is the infinite cyclic group.

46. Give the multiplication tables for $C_4 \times C_2$, $C_2 \times C_2 \times C_2$. Is either of these $\cong C_8$?

47. Prove that $C_5 \times C_2 \cong C_{10}$.

48. Give an independent set of generators and defining relations for $C_4 \times C_2$.

49. Repeat exercise 48 for $C_2 \times C_2 \times C_2$.

50. Show that the set of all numbers $x^m y^n$ (where x and y are *fixed* co-prime positive integers and m, n are any integers, positive, negative or zero) under multiplication, form a group $\cong$ the group of Gaussian integers.

Exercises 12B

1. By considering the element 3 show that the group of residues modulo 17 under multiplication is isomorphic to C_{16}, and find the orders and inverses of all the elements.

2. Find a generator of the group of residues modulo 23 under multiplication. Hence show that the group $\cong C_{22}$ and find an element of order 2 and one of order 11.

In 3–6 we denote the group of residues modulo n prime to n under multiplication by R_n.

3. Show that R_9 and R_{18} are both $\cong C_6$.

4. Show that $R_{15} \cong C_4 \times C_2$ under the isomorphism $2 \leftrightarrow (a, f)$ and $14 \leftrightarrow (e, b)$.

5. Find the orders of all the elements of R_{20} and express them in terms of $a = 3$, $b = 11$. Hence show that $R_{20} \cong C_4 \times C_2$.

6. Find the orders of all the elements of R_{24} and, by expressing them in terms of $a = 5$, $b = 7$, $c = 13$ show that $R_{24} \cong C_2 \times C_2 \times C_2$.

7. In D_{2n} denote the positions of the vertices of the n-gon by 1, 2, ..., n, taken clockwise. We define $x(i)$ to be the position into which the vertex at i is transformed by the transformation x, $x \in D_{2n}$. With the usual notation for the elements of D_{2n} prove

$$\text{(i) } a(i) \equiv i+1 \pmod{n};$$
$$\text{(ii) } b(i) \equiv 2-i \pmod{n}.$$

Hence prove that $(ab)(i) \equiv (ba^{n-1})(i) \equiv 1-i \pmod{n}$ and deduce that $ab = ba^{n-1}$.

8. Consider an infinite straight line with points marked along it at unit intervals. Let a be the transformation which moves it 1 unit to the right and let b rotate it through π about one of the marked points. Show that the group of transformations which leave the marked points occupying the same positions is generated by a and b with the defining relations (i) a has infinite order; (ii) $b^2 = e$; (iii) $aba = b$. Prove from (iii) that $a^r b a^r = b$ and verify this geometrically. Why is the group not $\cong Z \times C_2$, where Z is the infinite cyclic group?

9. If we have 2 regular tetrahedra $ABCD$ and $A'B'C'D'$ interlocking symmetrically (so that AB and $C'D'$ intersect and bisect each other, and similarly for the other pairs of sides) the resulting solid is called a *stella octangula*. Show that its transformation group is isomorphic to S_4.

10. A permutation $\begin{pmatrix} \lambda_1 & \lambda_2 \ldots \lambda_{r-1} & \lambda_r \\ \lambda_2 & \lambda_3 \ldots \lambda_r & \lambda_1 \end{pmatrix}$

which permutes $(\lambda_1\ \lambda_2 \ldots \lambda_r)$ cyclically is called a *cycle* and written

$(\lambda_1 \lambda_2 \ldots \lambda_r)$. Prove that any permutation of 1, 2, ..., n may be expressed as the product of cycles, the cycles permuting mutually exclusive subsets of the objects 1, 2, ..., n, and show that any 2 such cycles commute. (Such cycles are called *disjoint*.)

11. Expressed as a product of disjoint cycles we would write

$$\begin{pmatrix} 1 & 2 & 3 & 4 & 5 & 6 \\ 3 & 1 & 2 & 4 & 6 & 5 \end{pmatrix}$$

as (132)(56)(4). Express all permutations of S_3 and S_4 in this form.

12. Prove that the order of the cycle $(\lambda_1 \lambda_2 \ldots \lambda_r)$ is r, and that the order of any permutation is the L.C.M. of the orders of its component cycles, when it is expressed as a product of disjoint cycles.

13. Prove that the cycle $(\lambda_1 \lambda_2 \ldots \lambda_r)$ is even or odd according as r is odd or even and hence give a method of finding whether a given permutation is even or odd by expressing it as a product of disjoint cycles.

14. Let the set S consist of n members. Show that the group of subsets of S under $AB = (A \cup B) - (A \cap B)$ is isomorphic to $C_2 \times C_2 \times \ldots \times C_2$, where there are n terms.

15. In the group of subsets of S under $AB = (A \cup B) - (A \cap B)$ show that $ABCD$ is the subset consisting of all elements of S which are in 1 or 3 of the subsets A, B, C, D. Generalise this result for the product of n subsets.

16. If p and q are co-prime prove that $C_p \times C_q \cong C_{pq}$.

17. If p and q are not co-prime prove that the order of every element of $C_p \times C_q$ is a factor of the L.C.M. of p and q, and deduce that $C_p \times C_q$ is not isomorphic to C_{pq}.

18. Prove that there cannot be a finite set of generators for the group of rationals under addition.

19. If $a^2 = b^2 = e$ show that $ab = ba \leftrightarrow (ab)^2 = e$.

13

SUBGROUPS

13.1. Definition of a subgroup

When dealing with a set we are often interested in restricting our work to within a subset of it. When the set is a group we wish to apply our group theory to any subset with which we are working, and this cannot be done unless the subset is a group in its own right. A subset which is itself a group (under the same group product, of course) is called a *subgroup* of the given group.

When we work with a subgroup we are isolating the elements that we wish to consider, while retaining the group structure. As an example, if our original group is the complex numbers under addition, we may consider the subgroup of reals under addition, and even restrict our work to the subgroup of integers within this. We may, of course, consider a subgroup as existing independently: the usefulness of the idea lies largely in the fact that it shows us something of the structure of a group in terms of smaller, and often simpler, ones.

If H is a subset of a group G we already have a product defined between any two elements of H since they are also elements of G. For H to be a subgroup this product must be in the subset H for every pair of elements. The Associative Law holds since it does so in G, but we must have the neutral element in H and also the inverse of each element of H. (These latter must be the same in H as in G, since the neutral element and inverses in G act likewise in H and those of the latter are unique.) Hence we have the following definition.

Definition of a subgroup

A subset H of a group G is a subgroup of G if:

(*a*) Given any ordered pair (h_1, h_2) of elements of H the product h_1h_2 (as defined in G) is in H.

(*b*) The neutral element of G is in H.

(*c*) Given any element h in H the inverse h^{-1} (as defined in G) is in H.

Note 1. (*b*) is in fact superfluous, since if $h \in H$, $h^{-1} \in H$ by (*c*) and so $hh^{-1} = e \in H$ by (*a*).

Note 2. A subset which is not a subgroup may sometimes form a group under a different product. Thus the subset $\{2^n\}$ with n an integer is a group under multiplication, but is not a subgroup of the group of reals under addition. The structure as well as the elements needs to be part of the given group.

Theorem 13.1.1. *A necessary and sufficient condition for H to be a subgroup of G is that $gh^{-1} \in H$ for all $g, h \in H$.*

Necessary. If H is a subgroup $h^{-1} \in H$ and so $gh^{-1} \in H$.

Sufficient. If g is any element of H we have $gg^{-1} \in H$, i.e. $e \in H$. Hence $eg^{-1} = g^{-1} \in H$, i.e. H contains inverses. Thus if $g, h \in H$, so is h^{-1} and hence $g(h^{-1})^{-1} \in H$, i.e. $gh \in H$.

This theorem gives a useful practical and economical method of determining whether a given subset is a subgroup.

Trivial subgroups

The subset consisting of e alone is always a subgroup, as is the whole group G. These are called *trivial* subgroups: any others are *non-trivial.* All subgroups other than G are *proper.*

An infinite group may have finite subgroups (e.g. the subset $\{1, -1\}$ of the reals under multiplication), and a non-Abelian group may have an Abelian subgroup (e.g. the subset of reals of the quaternions under multiplication).

If K is a subgroup of H, which itself is a subgroup of G, then clearly K is a subgroup of G. We may likewise have a sequence of subgroups $G_1 \supset G_2 \supset G_3 \supset \ldots \supset G_n$ each contained in the preceding one.

13.2. The subgroup of powers of an element

If we look at the group of residues modulo 12 under addition, an obvious subgroup is formed by the set of the even residues, while another consists of all multiples of 3. But these two groups are the sets of all the powers of 2 and 3, respectively.

Similarly, in the dihedral group of order 10 (say), all the rotations which do not turn the pentagon over (i.e. all powers of

a in our notation in chapter 12) form a subgroup, as do the two elements e and b.

Similar subgroups may be identified in any group.

If g is any element of G the powers of g form a subset of G. This is always a subgroup, since $g^m . g^n = g^{m+n}$, and is in the set, $e = g^0$ is in it, as is $(g^n)^{-1} = g^{-n}$.

If g has infinite order the subgroup of its powers is infinite, while if the order of g is finite and equals n, the order of the subgroup of powers is also n. We see here a natural connection between the two uses of the word 'order'—as applied to elements and to groups.

The subgroup of powers of g is always Abelian and cyclic with g as a generator. We may obtain the same subgroup from different g's; in the group of residues modulo 8 under addition the subgroups of powers of 2 and 6 are the same. If g generates the whole group G then the subgroup is trivial, being G itself.

13.3. Examples of subgroups

There are many more or less obvious examples that could be given, and some will be met with later in the chapter. We restrict ourselves in this section to a few important types.

Subgroups of vectors

Considering the group of two-dimensional vectors under addition a subgroup is given by any subset consisting of all vectors of the form $(x, \lambda x)$ where λ is a fixed constant. This corresponds to a line through the origin. That such a subset is a subgroup may easily be verified, from the definition or by using theorem 13.1.1. A line not through the origin does not give a subgroup since it does not contain 0.

In three dimensions a line through the origin again gives a subgroup, which consists of all vectors of the form $(x, \lambda x, \mu x)$ where λ and μ are constants. In this case we also have a subgroup given by any plane through O, i.e. the set (x, y, z) with $\lambda x + \mu y + \nu z = 0$ for given λ, μ and ν. A line or plane not through O again does not form a subgroup. In fact, the sum of two vectors both in such a line or plane is *never* in it. Thus for a plane, if (x_1, y_1, z_1) and (x_2, y_2, z_2) are both in the subset given by $\lambda x + \mu y + \nu z = a$ for a non-zero, then

$$\lambda(x_1 + x_2) + \mu(y_1 + y_2) + \nu(z_1 + z_2) = 2a,$$

not a, and so the sum of the vectors is not in the set.

In n-dimensional vectors a subgroup is given by the vectors $(x_1, x_2, \ldots, x_n)$ which satisfy r independent equations $\sum_{i=1}^{n} \lambda_i^j x_i = 0$, $j = 1 \ldots r$ for any r where $1 \leqslant r \leqslant n-1$. This corresponds to an $(n-r)$-dimensional flat face through O.

Permutations leaving a subset of objects invariant

If in the group of permutations of n objects we select the subset of all those which leave certain specified objects unchanged then this forms a subgroup, for clearly if g and h are in the subset so is gh^{-1}. If r objects are unchanged this subgroup is isomorphic to S_{n-r}, since we may ignore the invariant objects and consider the subgroup as permutations on the remainder. Hence we have several subgroups of S_n all isomorphic to S_r for any $r < n$.

For example, with permutations on four objects those that leave the first unchanged are

$$\begin{pmatrix} 1 & 2 & 3 & 4 \\ 1 & 2 & 3 & 4 \end{pmatrix}, \begin{pmatrix} 1 & 2 & 3 & 4 \\ 1 & 3 & 4 & 2 \end{pmatrix}, \begin{pmatrix} 1 & 2 & 3 & 4 \\ 1 & 4 & 2 & 3 \end{pmatrix},$$

$$\begin{pmatrix} 1 & 2 & 3 & 4 \\ 1 & 2 & 4 & 3 \end{pmatrix}, \begin{pmatrix} 1 & 2 & 3 & 4 \\ 1 & 4 & 3 & 2 \end{pmatrix}, \begin{pmatrix} 1 & 2 & 3 & 4 \\ 1 & 3 & 2 & 4 \end{pmatrix}.$$

Similarly, there are 6 that leave the second object unchanged, and similarly for the other two objects. We obtain 4 subgroups all isomorphic to S_3. The isomorphism is seen most clearly in the case of those that leave the last object unchanged, these elements being identical to the elements of S_3 itself except for the addition of the symbol '4' to both rows: thus

$$\begin{pmatrix} 1 & 2 & 3 & 4 \\ 3 & 1 & 2 & 4 \end{pmatrix} \text{ corresponds to } \begin{pmatrix} 1 & 2 & 3 \\ 3 & 1 & 2 \end{pmatrix}.$$

The centre of a group

Consider the group of transformations of the equilateral triangle. This is non-commutative, but some pairs of elements do commute. Thus any two powers of a commute, and of course each element commutes with itself. The neutral element commutes with every element, of course, but no other element has this property—no power of a commutes with b, and none of b, ba, ba^2 commutes with a.

Similarly, in the permutation groups given any element x, except e, we can find some elements with which x does not commute (except in the permutation group of degree 2, which is commutative).

In the group of quaternions under multiplication (except 0), which is also non-Abelian, the non-commutativity enters only when we got outside the real numbers, which form part of the group: the reals themselves commute with all quaternions.

In any group we define the set of all those elements that commute with *all* elements of the group to be the *centre* of the group. It may be thought of as the Abelian part of the group.

If the group is Abelian, the centre coincides with the whole group. As we have seen from the examples given, in a non-Abelian group the centre may consist of e alone (it always includes e of course), or it may be a non-trivial subset of the elements.

The centre will be dealt with more fully in volume 2, but we will here prove that it is always a subgroup (not necessarily a proper subgroup of course). It is in fact an Abelian subgroup, since certainly any two of its elements must commute, since they commute with all elements of the original group.

Theorem 13.3.1. *The centre is a subgroup.*

Suppose g and h are in the centre. Then for any x in our group, $gx = xg$ and $hx = xh$. Hence $ghx = gxh = xgh$ and so gh is in the centre. For any x we have $gx^{-1} = x^{-1}g$ and so $(gx^{-1})^{-1} = (x^{-1}g)^{-1}$, i.e. $xg^{-1} = g^{-1}x$ and so g^{-1} is in the centre. Since $ex = xe$ $(= x)$ for any x we always have e in the centre. Thus the centre is a subgroup.

Subgroups generated by subsets of elements

Consider the group of three-dimensional vectors under addition, and take the subset (not subgroup) consisting of the two unit vectors (1, 0, 0) and (0, 1, 0), denoting these by $\mathbf{a}$ and

b. If we add these together repeatedly we will obtain new vectors, since **a** and **b** do not themselves form a subgroup. Thus $\mathbf{a}+\mathbf{b} = (1, 1, 0)$, $\mathbf{b}+\mathbf{a}+\mathbf{a}+\mathbf{a}+\mathbf{b}+\mathbf{b}+\mathbf{a} = (4, 3, 0)$, etc. Let us also allow the inverses of **a** and **b** in this addition process. Then we can easily see that however many additions we perform, using just **a** and **b** or their inverses, we will obtain just those vectors of the form $(r, s, 0)$ where r and s are integers, positive, negative or zero. Of course each of these will arise in many ways (e.g. $(1, 1, 0) = \mathbf{a}+\mathbf{b} = \mathbf{a}+\mathbf{b}+\mathbf{a}^{-1}+\mathbf{a}$, etc.), but all of them *will* arise, and no others. The vectors of the form given form a subset of the three-dimensional vectors which consists of all the points which are vertices of a grid of unit squares on the (x, y) plane. The important thing to note about them is that they form a subgroup—the sum of any two, the zero and the inverse (i.e. the negative) of each are all in the set.

Let us do the same with the subset of three vectors $\mathbf{a} = (1,0,0)$, $\mathbf{b} = (0, 1, 0)$ and $\mathbf{c} = (\frac{1}{2}, 0, 0)$. We then obtain all vectors of the form $(\frac{1}{2}r, s, 0)$ where r and s are integers, but there is still more choice in our additions, while obtaining the same answer. For example, $(1, 1, 0) = \mathbf{a}+\mathbf{b} = \mathbf{a}+\mathbf{b}+\mathbf{a}^{-1}+\mathbf{a} = \ldots$ as before, but it also equals $\mathbf{c}+\mathbf{c}+\mathbf{b}$, etc. However, the set we obtain still forms a subgroup.

Now consider the set of residues modulo 16 under addition. If we add together successive 2's, 4's or their inverses (which are 14 and 12) as in the above, we always obtain residues in the set 0, 2, 4, 6, 8, 10, 12, 14, i.e. all the even residues. Each may be obtained in many ways, but together they form a subgroup. If we do the same with the subset {4, 8} we obtain the subgroup of multiples of 4, but if we repeat with the subset {2, 3} we can obtain any residue in the whole group, and thus we have the trivial subgroup consisting of the group itself.

The above groups were commutative, but the same thing may be done with non-commutative groups. Consider the symmetric group of degree 4, and consider the subset comprising

$$a = \begin{pmatrix} 1 & 2 & 3 & 4 \\ 1 & 3 & 4 & 2 \end{pmatrix} \text{ and } b = \begin{pmatrix} 1 & 2 & 3 & 4 \\ 1 & 2 & 4 & 3 \end{pmatrix}.$$

Here $a^3 = e$ and $b^2 = e$, but we may build up complicated elements of the form, say, $abab^{-1}ab^{-1}a^{-1}b$ which, since a and b do not commute, may not be obviously simplified. But in all these cases the resulting element will not change the object 1, since neither a nor b does so. Thus all combinations will lie in the subgroup that leaves 1 invariant, and we can quite easily see that all such elements can arise, each in many ways of course. For example

$$\begin{pmatrix} 1 & 2 & 3 & 4 \\ 1 & 4 & 3 & 2 \end{pmatrix} = ab = ba^2, \text{etc.}$$

For another example, in the same permutation group let

$$a = \begin{pmatrix} 1 & 2 & 3 & 4 \\ 2 & 1 & 4 & 3 \end{pmatrix}, \quad b = \begin{pmatrix} 1 & 2 & 3 & 4 \\ 1 & 3 & 4 & 2 \end{pmatrix}.$$

Both these are even permutations, and hence by theorem 12.6.2 all products of them and their inverses are also even. We can soon verify that *all* even permutations may be obtained from them (e.g.

$$\begin{pmatrix} 1 & 2 & 3 & 4 \\ 2 & 3 & 1 & 4 \end{pmatrix} = bba, \quad \begin{pmatrix} 1 & 2 & 3 & 4 \\ 3 & 2 & 4 & 1 \end{pmatrix} = bab,$$

etc.), each in many ways of course, and thus the set of all combinations forms the alternating group of degree 4, again a subgroup of the original symmetric group.

We may repeat this work generally. If we have any subset of elements in a group we consider the set of all possible products formed by writing down finite strings of these elements or their inverses in any order, with repetitions if we wish. Thus considering the subset $\{g, h, k\}$ we would include elements such as $hk^{-1}k^{-1}gkkkh^{-1}g^{-1}$. If our elements commuted all such could be written in the form $g_1^{\lambda_1}g_2^{\lambda_2} \ldots g_r^{\lambda_r}$ where our subset consists of $g_1, g_2, \ldots, g_r$ and the λ_i's are positive, zero or negative integers; but for non-commutative elements more complicated expressions will arise. All products formed in the way described must be in the original group, though they will not all be different. The set of all such is a subset of the group, and is always a *subgroup*. We have seen this in all the special examples we gave, but it requires only a moment's thought to see that it is always so. For the product of two such elements is of the same form and so is in the set formed by all of them, the neutral element is in the set (it may be written as $g_1g_1^{-1}$ or in many other ways) and the inverse of any is of the same form (we merely take the set backwards and write down each inverse in turn—thus the inverse of

$$hk^{-1}k^{-1}gkkkh^{-1}g^{-1} \text{ is } ghk^{-1}k^{-1}k^{-1}g^{-1}kkh^{-1}).$$

We call this group the *subgroup generated by our subset.* It may be the whole group, as in the case of the subset {2, 3} in the residues modulo 16, in which case the subset is a system of generators for the whole group.

The subgroup generated by a subset is the smallest subgroup that contains that subset, for *any* subgroup which contains the subset necessarily contains all finite products of its elements and their inverses, and hence contains the subgroup generated by the subset. If we take *all* subgroups which contain the subset, then since they each contain the subgroup generated by the subset, so does their intersection. But this subgroup is itself one which contains the subset, and so the intersection must be precisely this subgroup. Thus in the group of residues modulo 60 under addition, the elements 12 and 36 generate the subgroup comprising all multiples of 12, and any subgroup which contains 12 and 36 includes this subgroup, examples being the subgroup of the even residues, that of all multiples of 3, that of multiples of 4 and that of multiples of 6.

Direct products

We recall the definition of direct product of two groups given in §12.9, and bear in mind the practical example given there of two-dimensional vectors, as the direct product of the group of the real numbers with itself. If we consider the vectors as represented in a plane, then we can clearly pick out the axes as subgroups, which are in effect identical with the groups of real numbers from which the direct product is formed. (We remember in the introduction of complex numbers that we eventually identified the reals with the complex numbers $(a, 0)$.)

In the case of the two-dimensional vectors, we have thought of their group as being the direct product of the two groups of reals represented by the co-ordinate axes, which have been taken as being rectangular. But we could use oblique axes, and still obtain the vectors as the direct product of two groups of reals. As an example, let us keep the same co-ordinates but consider the two subgroups A = the set of vectors of the form $(x, 0)$, B = the set (x, x). Geometrically A is the x-axis and B the line $y = x$. Then any vector is expressible uniquely in the form $\mathbf{a}+\mathbf{b}$, where $\mathbf{a} \in A$ and $\mathbf{b} \in B$. Thus $(-5, 7) = (-12, 0)+(7, 7)$ or, in terms of new co-ordinates, is $(-12, 7)$. The vectors here are given as the direct product of A and B, the form (a, b) being given in effect by the new co-ordinates.

Similarly, the three-dimensional vectors may be considered as the direct product of the group A of two-dimensional vectors formed by any plane through the origin, and of the group B of reals formed by any line through O, *provided B does not lie in the plane A*. If B were in A then a vector not in A could not be expressed in terms of those in A and B.

We will now endeavour to find conditions under which a group may be considered as the direct product of two or more of its subgroups. If such a decomposition is possible then we have obtained a picture of our group in terms of simpler ones (since proper subgroups of it are inevitably simpler, and the process of forming a direct product is intrinsically simple and is capable of straightforward investigation). Thus such an expression will simplify a complicated group considerably, and the method of expressing a given group as the direct product of two or more of its subgroups is a very powerful one for analysing the structure of groups.

Suppose first that G is of the required form, that is $G = A \times B$. It is easy to see that the set of elements of G given by (a, f), where f is the neutral element of B, form a subgroup isomorphic to A. [For

$$(a_1, f).(a_2, f) = (a_1 a_2, ff) = (a_1 a_2, f),$$
$$(a_1, f)^{-1} = (a_1^{-1}, f^{-1}) = (a_1^{-1}, f)].$$

We may identify this subgroup with A and say that A, and similarly B, is a subgroup of G. (Strictly speaking they are isomorphic to subgroups of G, in the same way that the reals are isomorphic to those complex numbers with zero imaginary part, but the identification causes no confusion in either case.)

Then (1) any element of G may be expressed in one and only one way as the product of an element of A with one of B, since the element

$$(a, b) = (a, f).(e, b)$$

and this factorisation is unique since, if $(a, b) = (x, f).(e, y)$ then $a = xe = x$ and $b = fy = y$ (e and f are the neutral elements of A and B respectively).

We also see (2) that any element of A commutes with any element of B, since

$$(a, f).(e, b) = (e, b).(a, f) = (a, b).$$

(Remember we are now dealing with general groups, not merely Abelian ones.)

We will now show that these conditions are sufficient for a group to be the direct product of two of its subgroups. Condition (1) is fairly obvious. We implied it when considering vectors: every vector had to be expressible as the product of two from the subgroups, and this expression was unique in every case. If it were not unique (for example if in the three-dimensional case A and B were both planes) then the group would not be the direct product, since the latter would contain too many elements, speaking loosely (if A and B are planes, their direct product is in four dimensions, not three). Thus condition (1) gives us a 1-1 correspondence that is necessary to prove the isomorphism of the given group with the direct product of the subgroups. Condition (2), of commutativity, did not enter into our discussion of vectors, where the whole group is commutative, but is necessary in the general case to ensure that the 1-1 correspondence preserves the group structure, i.e. to make sure that not only do G and $A \times B$ have the same elements, but also that they are the same *as groups*. The formal proof is given in the next theorem.

Theorem 13.3.2. *Suppose A and B are two subgroups of G with the properties:*

(*a*) *Every element of G is expressible uniquely in the form ab, $a \in A$, $b \in B$;*

(b) Every element of A commutes with every element of B. Then $G \cong A \times B$.

We set up a correspondence between the elements of G and $A \oplus B$ by $g \leftrightarrow (a, b)$, where a and b are the elements which give $g = ab$: such elements exist and are unique by (*a*).

We need to show that this correspondence is 1-1 and preserves product. Given g we know a and b uniquely and so (a, b) is unique. Given (a, b) we know $g = ab$ uniquely. Hence the correspondence is 1-1. If $g_1 = a_1 b_1$ and $g_2 = a_2 b_2$ then $g_1 g_2 = a_1 b_1 a_2 b_2 = a_1 a_2 b_1 b_2$ by (*b*).

Thus $g_1 g_2$ corresponds to $(a_1 a_2, b_1 b_2)$, which is the product of (a_1, b_1) and (a_2, b_2) in the group $A \times B$. Hence the correspondence preserves product and is an isomorphism.

The uniqueness in (*a*) above may be expressed in another way.

Theorem 13.3.3. *If A and B are subgroups of G and every element of G is expressible in the form ab for $a \in A$, $b \in B$, then this expression is unique if and only if $A \cap B = \{e\}$, the set consisting of the neutral element only.*

Suppose first that $A \cap B = \{e\}$, and let $g = ab = a_1 b_1$. Then $a_1^{-1} a = b_1 b^{-1}$ and so is in both A and B. Hence

$$a_1^{-1} a = b_1 b^{-1} = e,$$

i.e. $a = a_1$ and $b = b_1$. Thus the expression $g = ab$ is unique.

Conversely suppose $A \cap B \neq \{e\}$. Then there is an element x in A and B with $x \neq e$. If $g = ab$ we also have $g = axx^{-1}b$, i.e. $g = (ax).(x^{-1}b)$. But a and x are both in A and so ax is, while b, x and hence x^{-1} and $x^{-1}b$ are in B. Also $ax \neq a$ since $x \neq e$, and so the expression of g as a product of elements in A and B is not unique.

We may extend the above work to more than two subgroups. The result, which is proved similarly or by induction, is:

Theorem 13.3.4. *If G contains subgroups $A_1, A_2, \ldots, A_n$ which have the properties:*

(a) Every element of G is expressible uniquely in the form $a_1 a_2 \ldots a_n$, where $a_i \in A_i$,

(b) Every element of A_i commutes with every element of A_j for all $i \neq j$,

then $G \cong A_1 \times A_2 \times \ldots \times A_n$.

If A is defined in terms of the generators $\{a_i\}$ and defining relations $\{\Gamma_k\}$, and B by generators $\{b_j\}$ and relations $\{\Delta_l\}$, then a set of generators for $A \times B$ is given by the set of a_i's together with the b_j's, and a set of defining relations for $A \times B$ is given by all the Γ_k's and Δ_l's together with $a_i b_j = b_j a_i$ for all i, j.

13.4. Frobenius notation for subsets of a group

In this section we will extend the ordinary set notation, in the case where the sets are subsets of a group, so as to take account of the product structure. The notation introduced is perfectly natural and easy to use, but it is convenient to introduce it in an explicit manner, and we proceed to do this.

We will denote subsets of the set of elements of a group G by large letters, usually H, K, etc. (Single elements are usually referred to by small letters.) We then have the usual set ideas of $H \cup K$ for the set of elements in either H or K or both and $H \cap K$ for the set of elements in both H and K. $H = K$ means that they consist of the same elements, $H \supseteq K$ means that every element in K is in H, and $H \subseteq K$ that every one in H is in K, strict inclusions being denoted by $\supset$ or $\subset$.

In addition to the normal set structure we may take the product of any element of H with any of K. If we do this over all elements of H and K we obtain a collection of elements which is itself a subset of G, and we introduce the notation HK for this set. It may be called the *product* of H and K.

Definition. *HK means the set of all elements hk for $h \in H$ and $k \in K$.*

Note that the same element may be given by two or more different pairs hk.

As an example consider the two-dimensional vectors under addition. Let H be the vectors on the line $x = 0$, and K consist of the two elements $(1, 0)$ and $(2, 0)$. Then HK consists of the two lines $x = 1$ and $x = 2$. In this case each element of HK is given precisely once.

Now let H be the same and K consist of the three vectors $(1, 0)$, $(2, 0)$ and $(1, 1)$. Then HK is the same as before, but this time any vector on $x = 1$ can arise in two ways: thus

$$(1, 4) = (0, 4)+(1, 0) \quad \text{or} \quad (0, 3)+(1, 1).$$

Distinguish carefully between HK and $H \cup K$, and also between HK and $H \times K$. $H \cup K$ is the set of elements which are in either H or K or both—the product structure does not enter into it at all. In the first example above $H \cup K$ consists of the line $x = 0$ together with the two points $(1, 0)$ and $(2, 0)$, and bears little resemblance to HK containing points in fact that do not even belong to HK. $H \times K$ has no meaning unless H and K are both subgroups, and then indeed in some cases $H \times K$ may be the same as HK. For example, if H is the set of vectors on $x = 0$ and K is the set on $y = 0$ both $H \times K$ and HK are the whole group of two-dimensional vectors. But $H \times K$ *need not* equal HK. Thus in the case of three-dimensional vectors let H be those in the plane $x = 0$ and K be those in the plane $y = 0$. Then HK consists of the whole group of three-dimensional vectors. But $H \times K$ is a group in *four* dimensions, being in fact isomorphic to the group of four-dimensional vectors.

A subset of a group is often called a *complex*. Product of complexes is associative, since $H(KL) = (HK)L = \{hkl\}$, $h \in H$, $k \in K$, $l \in L$. It is not in general commutative, but may be in special cases. It is important to realise that $HK = KH$ does *not* mean that all pairs of elements of H and K commute, i.e. that $hk = kh$ for all $h \in H$ and $k \in K$: it merely expresses the fact that if we form all products hk we obtain the same elements as if we form products kh, but they need not be given in the same order.

Example. In the dihedral group of order 6 let $H = \{e, a, a^2\}$, $K = \{e, b\}$. Then $HK = KH = D_6$ but $ab \neq ba$, i.e. $hk \neq kh$ for all $h \in H$, $k \in K$.

One of H and K may be a single element g. We then write gH to mean the set of elements $\{gh\}$ as h varies over H, or Hg to be the set $\{hg\}$.

Extending the notation we write H^{-1} to mean the set $\{h^{-1}\}$ for $h \in H$. We also denote HH by H^2 with obvious extensions for higher powers.

For an example in two-dimensional vectors, if H is the set $x = 0$ and g is $(1, 0)$, gH is the set $x = 1$. If H is the set $x = 1$, H^{-1} is the set $x = -1$ (though $(1, y)$ gives rise to $(-1, -y)$ and not $(-1, y)$), while H^2 is the set $x = 2$.

We now wish to investigate the conditions for a complex to be a subgroup. The basic conditions for a subgroup are that the product of any two elements of the complex, the inverse of any element and the neutral element are all in the set. The fact that the product of any two elements is in the complex may be stated in the form $H^2 \subseteq H$, and we also need $H^{-1} \subseteq H$ and $e \in H$. The third condition follows from the first two, and we will show in fact that the inclusion signs may be replaced by equalities if H is a subgroup. Theorems 13.4.1 and 13.4.2 give the easy formal proofs. In theorem 13.4.3 we reduce the conditions in the case of finite groups to only just one inclusion condition. In terms of elements this condition is merely that the product of any two elements of the subset must be in that subset.

Theorem 13.4.1. *If H is a subgroup, $H^2 = H$ and $H^{-1} = H$.*

Since H is a subgroup, $hh' \in H$ if h and h' are, and so $H^2 \subseteq H$. But since $h = he$ and $e \in H$, $H \subseteq H^2$. Hence $H^2 = H$.

$h \in H \Rightarrow h^{-1} \in H$, therefore $H^{-1} \subseteq H$. Hence $(H^{-1})^{-1} \subseteq H^{-1}$, i.e. $H \subseteq H^{-1}$. Thus $H^{-1} = H$.

Theorem 13.4.2. *$H^2 = H$ and $H^{-1} = H$ together imply that H is a subgroup.*

If h, k are in H we have hk and h^{-1} are both in H, and so $hh^{-1} = e$ is in it. Hence H is a subgroup.

Theorem 13.4.3. *If G is finite $H^2 \subseteq H \Rightarrow H^{-1} \subseteq H$ and is sufficient to make H a subgroup.*

For any $h \in H$, $h^2 \in H^2$ and so $\in H$, $h^3 \in H$, etc. Thus if the order of h is n, $h^{-1} = h^{n-1} \in H$, i.e. $H^{-1} \subseteq H$. Thus for any $h, h' \in H$, $(h')^{-1} \in H$ and so $h(h')^{-1}$ is, making H a subgroup.

So we have the result that for a finite group G a necessary and sufficient condition for a complex H to be a subgroup of G is that $H^2 \subseteq H$.

Counter example for an infinite group

If H is infinite the proof in theorem 13.4.3 breaks down since we can no longer be sure that all the elements of H have finite order.

Let G be the group of reals under addition and take H to be the complex consisting of all *positive* integers. Then $H^2 \subseteq H$ since the sum of two positive integers is itself a positive integer, but H is not a subgroup as it contains neither unit nor inverses.

Theorem 13.4.4. *If H and K are subgroups so is* $H \cap K$.

If g and h are in $H \cap K$ they are both in H and both in K. Hence, since H and K are subgroups, gh^{-1} is in H and in K, and so in $H \cap K$. Hence $H \cap K$ is a subgroup by theorem 13.1.1.

Notice that $H \cup K$ need not be a subgroup, since if g and $h \in H \cup K$ there is no guarantee that they are both in either H or K, and hence we can say nothing about gh^{-1}.

We can show as in theorem 13.4.4 that the intersection of any number, finite or infinite, of subgroups is itself a subgroup.

Theorem 13.4.5. *If H and K are subgroups, HK is a subgroup if and only if* $HK = KH$.

If. If $HK = KH$, $(HK)^2 = HKHK = HHKK = HK$ by theorem 13.4.1 since H and K are subgroups. Also

$$(HK)^{-1} = K^{-1}H^{-1} = KH \quad \text{by theorem 13.4.1}$$
$$= HK.$$

Hence by theorem 13.4.2 HK is a subgroup.

Only if. If HK is a subgroup

$$\begin{aligned} HK &= (HK)^{-1} \quad \text{by theorem 13.4.1} \\ &= K^{-1}H^{-1} \\ &= KH \quad \text{by theorem 13.4.1 since } H \text{ and } K \text{ are subgroups.} \end{aligned}$$

13.5. Cayley's theorem

In the early work on groups, before their abstract nature was fully realised, the only type to be studied was that of the permutation groups, i.e. subgroups of the symmetric groups. Many of the results obtained are equally valid for all finite groups, and it can be shown that any finite group whatsoever is isomorphic to a permutation group, so that in a sense the early work embraces the whole theory for the finite case. The theorem which proves this is due to Cayley (1854) and is one of the most important in group theory, though simple to understand.

Theorem 13.5.1. *Cayley's theorem.*

Any finite group G is isomorphic to a subgroup of the symmetric group of degree n, where n is the order of G.

We consider the symmetric group as the group of permutations of the n elements of G, denoted by $g_1, g_2, \ldots, g_n$. If x is any element of G we consider the elements $g_1x, g_2x, \ldots, g_nx$. These are all different by the Cancellation Law and hence the mapping $g_i \leftrightarrow g_ix$ is a permutation of the elements, and so is an element of S_n.

Consider the correspondence

$$x \leftrightarrow \begin{pmatrix} g_1 \ldots g_n \\ g_1x \ldots g_nx \end{pmatrix}$$

between the elements of G and a subset of S_n.

It is clearly 1-1, since there is only one permutation corresponding to x, and if the permutation is given we know x, if it exists at all. (The last remark is necessary since only part of S_n is included in the correspondence. The isomorphism is between G and a *subgroup* of S_n; it is known technically as a monomorphism.)

The correspondence preserves the group structure, for xy corresponds to the permutation

$$\begin{pmatrix} g_1 \ldots g_n \\ g_1xy \ldots g_nxy \end{pmatrix},$$

which is the product of the permutations

$$\begin{pmatrix} g_1 \ldots g_n \\ g_1x \ldots g_nx \end{pmatrix} \quad \text{and} \quad \begin{pmatrix} g_1x \ldots g_nx \\ g_1xy \ldots g_nxy \end{pmatrix},$$

i.e. of

$$\begin{pmatrix} g_1 \ldots g_n \\ g_1x \ldots g_nx \end{pmatrix} \quad \text{and} \quad \begin{pmatrix} g_1 \ldots g_n \\ g_1y \ldots g_ny \end{pmatrix},$$

which correspond to x and y, respectively.

Hence the correspondence is an isomorphism between G and a subset of S_n, and so G is isomorphic to this subset, which must be a subgroup since it is isomorphic to the group G.

Let us take a simple example. Consider the Vierergruppe with the four elements e, a, b, c, also numbered 1, 2, 3, 4 in line with our usual notation for permutations. Then in the isomorphism we set up, e corresponds to the identity permutation

$$\begin{pmatrix} e & a & b & c \\ e & a & b & c \end{pmatrix}, \quad \text{or} \quad \begin{pmatrix} 1 & 2 & 3 & 4 \\ 1 & 2 & 3 & 4 \end{pmatrix}.$$

a corresponds to the permutation

$$\begin{pmatrix} e & a & b & c \\ ea & a^2 & ba & ca \end{pmatrix},$$

which is $\begin{pmatrix} e & a & b & c \\ a & e & c & b \end{pmatrix}$, or $\begin{pmatrix} 1 & 2 & 3 & 4 \\ 2 & 1 & 4 & 3 \end{pmatrix}$.

Similarly

$$b \leftrightarrow \begin{pmatrix} e & a & b & c \\ eb & ab & b^2 & cb \end{pmatrix} = \begin{pmatrix} e & a & b & c \\ b & c & e & a \end{pmatrix} \text{ or } \begin{pmatrix} 1 & 2 & 3 & 4 \\ 3 & 4 & 1 & 2 \end{pmatrix},$$

$$c \leftrightarrow \begin{pmatrix} e & a & b & c \\ ec & ac & bc & c^2 \end{pmatrix} = \begin{pmatrix} e & a & b & c \\ c & b & a & e \end{pmatrix} \text{ or } \begin{pmatrix} 1 & 2 & 3 & 4 \\ 4 & 3 & 2 & 1 \end{pmatrix}.$$

Thus the Vierergruppe is isomorphic to the subgroup of S_4 formed by the elements

$$\begin{pmatrix} 1 & 2 & 3 & 4 \\ 1 & 2 & 3 & 4 \end{pmatrix}, \quad \begin{pmatrix} 1 & 2 & 3 & 4 \\ 2 & 1 & 4 & 3 \end{pmatrix},$$

$$\begin{pmatrix} 1 & 2 & 3 & 4 \\ 3 & 4 & 1 & 2 \end{pmatrix}, \quad \begin{pmatrix} 1 & 2 & 3 & 4 \\ 4 & 3 & 2 & 1 \end{pmatrix}.$$

As an example of the isomorphism, $c = ab$ corresponds to

$$\begin{pmatrix} 1 & 2 & 3 & 4 \\ 4 & 3 & 2 & 1 \end{pmatrix} \quad \text{which is} \quad \begin{pmatrix} 1 & 2 & 3 & 4 \\ 2 & 1 & 4 & 3 \end{pmatrix} \begin{pmatrix} 1 & 2 & 3 & 4 \\ 3 & 4 & 1 & 2 \end{pmatrix}.$$

Cayley's theorem gives one isomorphism between G and a subgroup of S_n. It need not be unique and in particular, since S_n is given as a subgroup of S_m $(m > n)$, by keeping $(m-n)$ objects invariant, G is isomorphic to subgroups of S_m. It often happens that G is isomorphic to a subgroup of S_r for $r < n$: any subgroup of S_r of order more than r is such a group G.

The importance of Cayley's theorem is largely theoretical; from the practical viewpoint it has little use. Except for the smallest groups it gives a subgroup of a needlessly large and complicated symmetric group. Thus a group of order 8 is given as a subgroup of one of order 8! or 40,320.

13.6. Cosets

In the group of two-dimensional vectors a line through the origin forms a subgroup, but a line not through O does not. Such a line obviously bears some resemblance to one through O, particularly to its parallel, and all lines parallel to a given one, Γ, through O form a decomposition of the whole group of vectors into subsets similar to Γ though not subgroups. The condition for two vectors $\mathbf{a}$ and $\mathbf{b}$ to lie on the same line is that $(\mathbf{a}-\mathbf{b})$ is a vector on Γ. We have an analogous idea in vectors of three or more dimensions. In three dimensions we may decompose the whole group either into lines parallel to a line Γ through O, or into planes parallel to a plane π through O, and in either case the condition for two vectors to belong to the same element of the decomposition is that their difference is a vector of Γ or π.

A similar idea is seen in the group of integers under addition. The set of integers divisible by n forms a subgroup H. Now consider all integers which have remainder r when divided by n. This set is in some respects similar to H—it consists of the elements of H plus r and is in a sense 'parallel' to H. It is not a subgroup. As r takes successively the values $0, 1, \ldots, (n-1)$ we obtain a decomposition of the whole group of integers into a collection of these 'parallel' subsets, and the condition that two integers belong to the same subset is that their difference is divisible by n, i.e. is in H.

In both the above examples we took a subgroup and formed a decomposition of our whole group into subsets 'parallel' to H, two elements being in the same subset if their difference was in H. The idea may be generalised to any group and subgroup of it. The 'difference' in the product notation becomes quotient of two elements g and k, which may be written either as gk^{-1} or $k^{-1}g$. We may take either of these, but will find it convenient to use the latter. Thus we will decompose our group G into subsets such that g and k are in the same one if $k^{-1}g \in H$. The subsets are known as *cosets*. To prove that we always have a true decomposition we use the idea of equivalence relation introduced in §2.4. The reader is advised to read that section again before proceeding.

Thus if G is any group with H a subgroup we say that gRk if $k^{-1}g \in H$.

Theorem 13.6.1. *The relation defined above is an equivalence relation.*

Reflexive. We always have gRg since $g^{-1}g = e \in H$ since H is a subgroup.

Symmetric. If gRk then $k^{-1}g \in H$. Hence its inverse is in H, i.e. $g^{-1}k \in H$ and so kRg.

Transitive. If gRk and kRl then $k^{-1}g \in H$ and $l^{-1}k \in H$. Hence $(l^{-1}k)(k^{-1}g) = l^{-1}g \in H$ and so gRl.

Note in the above that we have used precisely those properties of H which make it a subgroup—for the work we are doing at the present it is vital that H is a subgroup.

By the equivalence classes theorem (theorem 2.4.1) we have a decomposition of G into a set of equivalence classes, mutually exclusive, such that two elements g and k are in the same class if and only if $k^{-1}g \in H$. These classes are called the *left cosets of G relative to H.*

We may consider cosets from a slightly different point of view. The coset containing a given element g, i.e. the equivalence class containing g, consists of those elements k such that gRk, i.e. $k^{-1}g \in H$. But this is true if and only if $g^{-1}k \in H$, i.e. $k = gh$ for some $h \in H$. Hence the coset containing g is precisely the complex gH, the set $\{gh\}$ for $h \in H$. We thus have the following theorem.

Theorem 13.6.2. *The left cosets of G relative to H are the complexes gH, $g \in G$. Any left coset may be expressed in this form for any g in it. Any two cosets are either the same or have no element in common.*

We may similarly define *right cosets* of G relative to H as the equivalence classes under the relation given by $gk^{-1} \in H$. They are the complexes Hg.

Although in general, for non-Abelian groups, left and right cosets do not coincide their properties are the same and either may be used. It is advisable to restrict oneself to either one or the other, and so we will usually use left cosets, and will call them merely *cosets.*

We could *define* a coset as gH and deduce that this gave a decomposition into mutally exclusive classes, but our method is simpler and better, provided the idea of equivalence relationship is understood. The condition that $k^{-1}g \in H$ may equally well be given as $g^{-1}k \in H$. (But not of course as $gk^{-1} \in H$ or $kg^{-1} \in H$, since these lead to right cosets.)

Theorem 13.6.3. *The only coset which is a subgroup is H itself.*

H is a coset since it is of the form hH for any h in H. If gH is a subgroup it contains e and so may be written eH, which is H.

We have said that right and left cosets are not necessarily the same. They are for Abelian groups, and in certain other cases. A subgroup H which has its left and right cosets in G identical is called an *invariant subgroup of G*. Invariant subgroups are an extremely important type and form the basis of the more advanced parts of the theory of groups. They will be studied in volume 2.

As an example of a subgroup which is not invariant consider the dihedral group of order 6. The subset $\{e, b\}$ is a subgroup H. $aH = \{a, ab\}$ while $Ha = \{a, ba\}$ and these are not the same, since $ab \neq ba$.

To return to our examples in vectors and residues, we see that a coset of two-dimensional vectors relative to a line Γ through O consists of the set $\mathbf{a}+\mathbf{x}$ where $\mathbf{x} \in \Gamma$. For residues a coset relative to the subgroup of multiples of n consists of the integers $(r+\alpha n)$ for some r. Although we may think of r as between 0 and $n-1$ this is not necessary, and in general no one element of a coset is more important than another.

Other examples of cosets

Example 1. In the group of reals under addition consider the subgroup of integers. Then the coset of any real number x relative to this subgroup is the set $\{x+r\}$ for r an integer. Thus the coset containing $\frac{1}{2}$ consists of $\pm\frac{1}{2}$, $\pm 1\frac{1}{2}$, $\pm 2\frac{1}{2}$, etc.

Example 2. In the direct product $G \times H$ consider the subgroup H. The coset containing (g, h) is the subset $(g, h+H)$, i.e. the subset (g, H).

Example 3. In the dihedral group consider the subgroup generated by a, i.e. the subgroup of all elements which do not turn the polygon over. There are just two cosets, one being the subgroup itself and the other consisting of all other elements, i.e. all those which do turn the polygon over.

13.7. Lagrange's theorem

The reason for the importance of the notion of cosets, and for our introduction of it, is two-fold. In the first place cosets form a necessary introduction to the deeper theory of groups based on the idea of an invariant subgroup and its factor group, which are to be discussed in volume 2. More important for our present purposes, the study of cosets leads immediately to Lagrange's Theorem, which is one of the most fundamental in the whole subject.

As we have already implied, the study of a group becomes largely the study of its subgroups. To discover all subgroups of a given group is not a simple task, but we can limit our search in the finite case by applying Lagrange's theorem.

Theorem 13.7.1. *Lagrange's theorem.*

The order of any subgroup of a finite group is a factor of the order of the group.

Suppose H is a subgroup of G, where the order of G is n and that of H is m. Then we wish to prove that m is a factor of n.

The whole group G may be decomposed into cosets gH relative to H. The number of elements in each coset must be m, since the coset gH is formed by taking all elements of H and pre-multiplying them by g, and all the elements so formed are different by the Cancellation Law. But the cosets are completely distinct. Hence if there are r cosets we have that $n = m.r$, and so m is a factor of n.

The number r is called the *index* of H in G.

Theorem 13.7.2. *The order of any element of a finite group is a factor of the order of the group.*

Consider the subgroup of powers of the element. Its order is equal to that of the element, and the result follows by Lagrange's theorem.

Theorem 13.7.3. *If x is any element of a finite group G, where G is of order n, $x^n = e$.*

Let x have order m. Then $n = m.r$ and $x^m = e$. Hence $x^n = (x^m)^r = e$.

Theorem 13.7.4. The only group of order p, where p is prime, is the cyclic group of order p.

Take any non-neutral element x. The order of x must be a factor of p, by theorem 13.7.2. But p is prime, and so x has order p. Thus the subgroup generated by x must be the whole group, which is therefore cyclic.

Note that we have proved the fairly obvious fact that any non-neutral element is a generator.

13.8. Subgroups of cyclic groups

Theorem 13.8.1. The only subgroups of C_p, where p is prime, are the trivial ones $\{e\}$ and C_p.

By Lagrange's theorem the order of a subgroup must be either 1 or p since p is prime.

For cyclic groups of composite order n we know that any subgroup must have order m where m is a factor of n. We will show that a unique subgroup corresponding to each factor exists.

Theorem 13.8.2. Corresponding to each factor m of n there exists a unique subgroup of C_n, and all these subgroups are cyclic.

Let $n = m.r$. To show that such a subgroup exists we consider the elements $e, g^r, g^{2r} \dots g^{(m-1)r}$ where g is a generator of C_n. These are distinct and clearly form a subgroup, which has order m and is cyclic, being generated by g^r.

Now suppose that we are given a subgroup of order m. Any element x of it is such that $x^m = e$, by theorem 13.7.3. Let $x = g^\lambda$ where g is a generator of C_n. Then $x^m = g^{m\lambda} = e$, and so $m\lambda$ must be a multiple of n, say $m\lambda = kn = kmr$. Hence $\lambda = kr$. Thus all elements of the subgroup are of the form g^{kr}, and since there are only m of these the subgroup must consist precisely of them, i.e. is that generated by g^r. Hence there is only one subgroup of order m.

Theorem 13.8.3. All non-trivial subgroups of the infinite cyclic group are themselves infinite cyclic groups.

The infinite cyclic group is isomorphic to the group of integers under addition, and so we may identify it with this. Let the smallest positive integer in a non-trivial subgroup be r. Then all numbers 0, $\pm kr$, for any positive integer k, are in the subgroup, and no other number is. (For suppose $s = kr+t$ for $1 \leqslant t \leqslant r-1$ is in the subgroup. Then $s-kr (= t)$ is in it, which would give a positive integer smaller than r.) Thus the subgroup is the infinite cyclic group generated by r, and there is one corresponding to each positive integer.

Theorem 13.8.4. *If a group G has no non-trivial subgroups it is isomorphic to C_p, where p is a prime.*

Take any non-neutral element x and consider the subgroup of powers of x. Since G has no non-trivial subgroups this must be the whole group G, i.e. G is generated by x, and so is cyclic. But both the infinite cyclic group and C_n for n composite possess proper subgroups by our previous theorems. Hence $G \cong C_p$ for p prime.

13.9. Classification of groups to order 6

There is no general method of finding all groups of a given order. We have already seen that when p is prime the only group of order p is the cyclic group. In certain other special cases it may be quite easy to find all the groups having a given order, but often the problem is extremely difficult, and all possibilities may not be known, especially for an order which possesses many factors (e.g. 60).

For very small orders it is possible to use *ad hoc* methods, which we proceed to do for numbers up to 6. This process rapidly becomes impossibly tedious.

The only groups of order 1, 2, 3 or 5 are the cyclic groups, by theorem 13.7.4.

Theorem 13.9.1. *There are just two groups of order* 4, *the cyclic group and the Vierergruppe* ($C_2 \times C_2$), *both Abelian.*

(*a*) Suppose that there exists an element of order 4. Then the group is cyclic, with this element as a generator.

(*b*) Suppose all elements, except e, have order 2. (All must have order 1, 2 or 4 by theorem 13.7.2.) Let two non-neutral elements be a and b, so that $a^2 = b^2 = e$. Consider ab. If $ab = e$, $a = b^{-1} = b$, which is not true. If $ab = a$, $b = e$, while if $ab = b$, $a = e$. Hence ab must be the fourth element, and similarly ba must be the same, and so the group is the Vierergruppe.

Theorem 13.9.2. *There are just two groups of order* 6, *the cyclic group and the dihedral group. The latter is non-Abelian and so is the smallest non-Abelian group.*

(*a*) Suppose there is an element of order 6. Then the group is the cyclic group, with this element as a generator.

(*b*) Suppose no element has order 6, but that there exists at least one of order 3.

If a has order 3, the 3 elements e, a, a^2 are distinct. Let one of the remaining elements be b, and consider the 3 elements b, ba, ba^2. These are all distinct, since e, a, a^2 are. None of them, furthermore, is a power of a. For suppose $ba = a$. Then $b = e$. Similarly, $ba = a^2$ implies $b = a$, and $ba = e$ implies $b = a^2$. Similarly, we cannot have ba^2 a power of a. Thus we have 6 distinct elements e, a, a^2, b, ba, ba^2 and so the group is composed of these 6 elements. Now consider b^2. It is an element of the group and is not b, ba or ba^2 since this would give b a power of a. If $b^2 = a$ we have $b^3 = ba$, $b^4 = a^2$, $b^5 = ba^2$, $b^6 = a^3 = e$. Hence b has order 6, which is contrary to the hypothesis of (*b*). Similarly if $b^2 = a^2$. Hence $b^2 = e$.

Now consider ab. ab cannot be a power of a since b is not. If $ab = b$ we would have $a = e$. If $ab = ba$ we have

$$(ab)^2 = a^2b^2 = a^2,\ (ab)^3 = a^3b^3 = b,\ (ab)^4 = a,$$
$$(ab)^5 = a^2b = ba^2,\ (ab)^6 = e,$$

and so ab would have order 6, which would give us (*a*). Hence $ab = ba^2$ and we have the group generated by the elements a and b with $a^3 = b^2 = e$ and $ab = ba^2$, i.e. we have the dihedral group.

(*c*) Suppose no element has order 6 or 3. Then all non-neutral ones have order 2. Let two of these be a and b. Then ab cannot be e, a or b as in the proof of theorem 13.9.1 (*b*). Hence ab is a fourth element. Consider ba. Since it has order 2 we must have $ba = (ba)^{-1} = a^{-1}b^{-1} = ab$, i.e. a and b commute. Hence the generators a and b give us four elements only, since they have order 2, viz. e, a, b, ab. These 4 elements form a subgroup of order 4 (viz. $C_2 \times C_2$), which contradicts Lagrange's theorem, since 4 is not a factor of 6. Thus this case does not give us a group.

We will not pursue this method to higher orders, but will give the result for the next composite number, 8. A proof will be found in Ledermann (*Introduction to the Theory of Finite Groups*).

Theorem 13.9.3. *There are five distinct groups of order* 8, *three Abelian and two non-Abelian.*

(*a*) C_8. Generator a with $a^8 = e$ (Abelian).

(*b*) $C_4 \times C_2$. Generators a and b with $a^4 = b^2 = e$ and $ab = ba$ (Abelian).

(*c*) $C_2 \times C_2 \times C_2$. Generators a, b and c with

$$a^2 = b^2 = c^2 = e \quad \text{and} \quad ab = ba, \quad ac = ca, \quad bc = cb$$

(Abelian).

(*d*) The dihedral group. Generators a and b with

$$a^4 = b^2 = e \quad \text{and} \quad ab = ba^3$$

(non-Abelian).

(*e*) The quaternion group. Generators a and b with $a^4 = e$, $a^2 = b^2$ and $ab = ba^3$ (non-Abelian).

Worked exercises

1. List all subgroups of $C_2 \times C_2 \times C_2$.

Let the generators be a, b, c. The order of any subgroup other than the trivial ones must be 2 or 4, since it is a factor of 8, the order of the whole group. The subgroups of order 2 must consist of e and an element of order 2. But every element has this order, and so we obtain seven subgroups each consisting of e and one other element.

Consider the subgroups of order 4. Since every element has order 2 all such subgroups must be isomorphic to the Vierergruppe, and it is easy to see that there are just seven such, generated by a, b; a, c; b, c; a, bc; b, ca; c, ab; ab, bc.

2. Give the left and right cosets of D_{12} relative to the subgroup

$$H = \{e, a^3\}.$$

The left cosets are

$$H = \{e, a^3\},$$
$$aH = \{a, a^4\},$$
$$a^2H = \{a^2, a^5\},$$
$$bH = \{b, ba^3\},$$
$$baH = \{ba, ba^4\},$$
$$ba^2H = \{ba^2, ba^5\}.$$

The right cosets are

$$H = \{e, a^3\},$$
$$Ha = \{a, a^4\},$$
$$Ha^2 = \{a^2, a^5\},$$
$$Hb = \{b, a^3b\} \quad \text{or} \quad \{b, ba^3\},$$
$$Hba = \{ba, a^3ba\} \quad \text{or} \quad \{ba, ba^4\},$$
$$Hba^2 = \{ba^2, a^3ba^2\} \quad \text{or} \quad \{ba^2, ba^5\}.$$

Note that the left and right cosets coincide in this case.

3. Find the centre of the quaternion group of order 8 defined by generators a, b with $a^4 = e$, $a^2 = b^2$ and $ab = ba^3$.

The elements are $e, a, a^2, a^3, b, ba, ba^2, ba^3$. Since $ab \neq ba$ and $a^3b = ba^9 = ba \neq ba^3$ we cannot have a, a^3 or b in the centre. $ba.a = ba^2$ and $a.ba = ba^4 = b$, so ba is not in the centre. Similarly, it is easy to see that neither ba^2 or ba^3 is in the centre. Consider a^2. It commutes with all powers of a, and $a^2b = ba^6 = ba^2$, $a^2.ba = ba^2.a = ba.a^2$, and similarly $a^2.ba^2 = ba^2.a^2$ and $a^2.ba^3 = ba^3.a^2$. Hence the centre is the subgroup $\{e, a^2\}$.

Exercises 13A

Which of the subsets in **1–8** are subgroups?

1. The subset of integers in the group of reals under addition.

2. The subset of integers in the group of reals except 0 under multiplication.

3. All numbers of the form $kn+1$, n a fixed integer, in the group of integers.

4. All vectors of the form either $(x, \lambda x)$ or $(x, \mu x)$, where λ and μ are fixed and $\lambda \neq \mu$, in the group of two-dimensional vectors.

5. All polynomials of degree at most 2 in the group of polynomials under addition.

6. All quadratic polynomials with unit coefficient of x^2 in the group of polynomials under addition.

7. All transformations of a cube which leave a given vertex fixed.

8. All transformations of a cube which leave one, or both, of two given vertices fixed.

List all the subgroups of the groups in **9–14** and give their indices.

9. The Vierergruppe.

10. C_{12}.

11. D_6.

12. $C_4 \times C_2$.

13. $C_3 \times C_2$.

14. The alternating group of degree 4. (The tetrahedral group.)

Find the centres of the groups in **15–18**.

15. D_{2n}.

16. S_n.

17. The group of quaternions under multiplication.

18. The group on page 247 defined by $(a, b).(c, d) = (ac, bc+d)$.

19. Prove that $H \cup K$ is a subgroup if and only if either $H \subseteq K$ or $K \subseteq H$, where H and K are subgroups of a group G.

20. By considering D_6 show that there may exist an Abelian subgroup which is larger than the centre.

21. Show that if H is a subgroup of G, then $\{g^{-1}Hg\}$, for any $g \in G$, is a subgroup isomorphic to H but not necessarily distinct from it.

22. Prove that $H(K \cup L) = HK \cup HL$ for any complexes H, K, L of G.

23. Prove that $H(K \cap L) \subseteq HK \cap HL$. Why do we not have equality? Give an example where $H(K \cap L) \subset HK \cap HL$.

24. Give an isomorphism between C_n and a subgroup of S_n.

25. Show that the correspondence

$$x \leftrightarrow \begin{pmatrix} q_1 & \dots & q_n \\ x^{-1}q_1 & \dots & x^{-1}q_n \end{pmatrix}$$

gives an isomorphism between the elements of G (order n) and a subset of S_n.

Give the left and right cosets of the groups in **26–31** relative to the given subgroups.

26. The Vierergruppe relative to $\{e, a\}$.

27. C_{12}, generator a, relative to $\{e, a^4, a^8\}$.

28. The complex numbers under multiplication relative to the reals.

29. The polynomials of degree at most 2 under addition relative to the subgroup of those with zero coefficient of x^2.

30. S_4 relative to the subgroup.

$$\left\{\begin{pmatrix} 1 & 2 & 3 & 4 \\ 1 & 2 & 3 & 4 \end{pmatrix}, \begin{pmatrix} 1 & 2 & 3 & 4 \\ 2 & 1 & 4 & 3 \end{pmatrix}, \begin{pmatrix} 1 & 2 & 3 & 4 \\ 3 & 4 & 1 & 2 \end{pmatrix}, \begin{pmatrix} 1 & 2 & 3 & 4 \\ 4 & 3 & 2 & 1 \end{pmatrix}\right\}.$$

31. D_6 relative to $\{e, b\}$.

32. Is $C_3 \times C_3 \cong C_9$? Explain your answer.

33. Show by giving a counter example that if H, K, L are complexes of a group G, $HK = HL$ does not necessarily imply that $K = L$. Prove that $gK = gL \Rightarrow K = L$.

34. If H is a subgroup and K a complex with $K \subseteq H$, show that $HK = H = KH$. Does this result hold if H is not a subgroup?

35. If H is a finite complex show that $H^2 \subseteq H \Rightarrow H^2 = H$ and hence that H is a subgroup.

36. Prove that if H and K are subgroups whose orders are r and s where r and s are co-prime then $H \cap K = \{e\}$.

37. Show that the group of the polynomial $x_1 x_2 + x_3 + x_4$ is a subgroup of the group of $x_1 x_2 + x_3 x_4$.

38. Prove that $(H \cap K)(H \cap L) \subseteq H \cap KL$ for any subgroup H and complexes K, L of G.

39. Prove that $(H \cup K)(H \cup L) \supseteq H \cup KL$ for any subgroup H and complexes K, L of G.

Exercises 13B

1. Show that in the group of two-dimensional vectors with integer coefficients under addition the only non-trivial subgroups are the sets $(\lambda rt, \mu rt)$ for λ, μ, t fixed and r any integer, positive, negative or zero.

2. Show that the only elements of finite order in the group of complex numbers under multiplication are those of the form $e^{2\pi ki/n}$ and hence show that the only finite subgroups are cyclic, generated by $e^{2\pi i/n}$ for n any positive integer.

3. Set up an isomorphism between the group of cross-ratios and a subgroup of S_4.

4. Show that the group of two-dimensional vectors may be expressed as the direct product of *any* two distinct lines through O, and that the group of three-dimensional vectors may be expressed either as the direct product of any three non-coplanar lines through O, or as the direct product of any plane through O and any line through O not in that plane.

5. By using the cycle notation prove that S_n has at least $(n-1)!$ elements of order n, and at least $n!/[(n-r)!\,r]$ of order $r < n$.

6. Show that the only two groups of order 9 are C_9 and $C_3 \times C_3$.

7. Show that the only two groups of order 10 are C_{10} and D_{10}.

8. Give the multiplication table for the quaternion group generated by a and b with $a^4 = e$, $a^2 = b^2$ and $ab = ba^3$.

9. By use of Lagrange's theorem and its corollaries prove Fermat's theorem that if p is prime and x is prime to p then $x^{p-1} \equiv 1 \pmod{p}$. Prove further that if n is not prime, but x is prime to n, then $x^{\phi(n)} \equiv 1 \pmod{n}$ where $\phi(n)$ is Euler's function (the number of numbers less than n prime to n).

10. If A_i ($i = 1, \ldots, r$), are subgroups of a group G, show that $B = \bigcap_{i=1}^{r} A_i$ is the greatest common subgroup of $A_1, \ldots, A_r$ in the following sense:

(i) B is a subgroup of each A_i, proper or improper but not empty;

(ii) any common subgroup of the A_i's is a subgroup of B. [Note the analogy with H.C.F.]

11. In the group of integers let H_r denote the subgroup generated by r, where r is a positive integer. Prove

(i) $H_r H_s$ is generated by r and s and is H_λ where λ is the H.C.F. of r and s.

(ii) $H_r \cap H_s = H_\mu$ where μ is the L.C.M. of r and s.

State and prove analogous results in the cyclic group of order n, and in this case show further that the index of H_λ is the H.C.F. of the indices of H_r and H_s, while the index of H_μ is their L.C.M.

12. In an Abelian group G let the subgroups H and K be generated by the sets of elements $\{h_1, \ldots, h_r\}$ and $\{k_1, \ldots, k_s\}$, respectively. Show that HK is a subgroup and is generated by $\{h_1, \ldots, h_r, k_1, \ldots, k_s\}$, which need not be independent.

13. Suppose H, K are subgroups of a group G, of orders r and s, respectively. Let $L = H \cap K$ have order d. Prove that if the cosets of H relative to L are $h_1L, \ldots, h_\lambda L$ then $HK = h_1K \cup \ldots \cup h_\lambda K$. Show further that no two of the h_iK's have elements in common and deduce that the number of elements in the complex HK is rs/d. [This is analogous to theorem 4.7.1.]

ANSWERS TO EXERCISES

Exercises 2A (p. 21)

26. True. **27.** False. **28.** True. **29.** False. **30.** True.

31. First part false, second true. **32.** True. **33.** True.

34. False.

42. No. **43.** Yes. **44.** No. **45.** No. **46.** Yes.

47. Yes. **48.** Yes. **49.** No. **50.** No.

Exercises 3A (p. 47)

14. $\varnothing, A, A, \varnothing, \varnothing$.

19. Nos. 1, 2, 3, 7, are self-dual, with possibly an interchange of the sets involved.

27. All my potatoes in this dish are old ones.

28. No jug in this cupboard will hold water.

29. Puppies that will not lie still never care to do worsted-work.

30. No M.P. should ride in a donkey race, unless he has perfect self-command.

31. None but red-haired boys learn Greek in this school.

32. An egg of the Great Auk is not to be had for a song.

33. No kitten with green eyes will play with a gorilla.

Exercises 3B (p. 49)

12. (i) $\bar{A} \cap (B \cup C)$; (ii) $(A \cap B) \cup (\bar{A} \cap \bar{B})$; (iii) $\bar{A} \cap (\bar{B} \cup (\bar{C} \cap \bar{D}))$.

15. 350. **16.** 18.

Exercises 4A (p. 69)

14. 13. **15.** 24. **16.** 7.

17. 1. **18.** 183. **19.** 1.

20. H.C.F. = 2, L.C.M. = $2^4.5^2.7$.

21. H.C.F. = 13, L.C.M. = $2^2.3.5.7.11.13$.

22. H.C.F. = 7, L.C.M. = $2.3.5.7^3.17.37$.

23. Only 0. All integers.

Exercises 4B (p. 70)

2. n^2-n+2, $\frac{1}{6}(n+1)(n^2-n+6)$.

4. All odd integers from 1 to $2^{n+1}-1$.

Exercises 5A (p. 83)

15. $x^4-10x^2+1 = 0$.

23. $xy = 5i+j+2k$, $yx = 5i-j+2k$, $xyz = 6+7i-j-2k$,
$zxy = 6+3i+3j+6k$, $zyx = 4+3i+j+8k$.

Exercises 6A (p. 102)

2. 1, 2 become 0, 1.

3. 1, 2, 3, 4, 5, 6 become 0, 2, 1, 4, 5, 3 (or 0, 4, 5, 2, 1, 3).

21. 5. 22. 8. 23. 15. 24. 27. 25. 311. 26. 2, 11, 20.

27. No solution. 28. 17, 46, 75, 104, 133.

29. No solution.

30. $2+3k$, $k = 0, 1, \ldots, 112$.

31. 5 or 14. 32. 6 or 11. 33. (i) 3; (ii) 5, 6.

Exercises 6B (p. 103)

3. $p = 3:2$; $p = 5:2, 3$; $p = 7:3, 5$; $p = 11:2, 6, 7, 8$.

7. 6, 28, 496.

Exercises 7A (p. 126)

1. All those of degree 0.

2. $2x^2+3x-3$, 5.

3. $x^5+x^4-x^2-x$, x.

4. $\frac{1}{3}x^3+\frac{1}{3}x^2+1$, $2x+3$.

5. $2x^4-8x^3+32x^2-128x+513\frac{1}{3}$, $-513\frac{1}{3}x-2$.

6. $x^{10}+x^8+x^6+x^4+x^2+1$, 0.

7. x^2-1. 8. x^2+3x+1. 9. 1.

10. x^6+x^5-2x-2. 11. $x^2+\frac{1}{2}$. 12. $x(x-1)(2x-1)$.

13. $(x-\sqrt[3]{2})(x-\omega\sqrt[3]{2})(x-\omega^2.\sqrt[3]{2})$, $(x-\sqrt[3]{2})(x^2+\sqrt[3]{2}x+2^{2/3})$, x^3-2.

14. $(x+1)(x-1)(x+i)(x-i)$, $(x+1)(x-1)(x^2+1)$.

15. $(x+5)(x+\frac{3}{2}+\frac{1}{2}\sqrt{5})(x+\frac{3}{2}-\frac{1}{2}\sqrt{5})$, $(x+5)(x^2+3x+1)$.

16. $(2x-3)(x+i)(x-i)$, $(2x-3)(x^2+1)$.

17. $(x-\frac{1}{2}+\frac{1}{2}\sqrt{5})(x-\frac{1}{2}-\frac{1}{2}\sqrt{5})(x+\sqrt{2}\,i)(x-\sqrt{2}\,i)$,
$(x-\frac{1}{2}+\frac{1}{2}\sqrt{5})(x-\frac{1}{2}-\frac{1}{2}\sqrt{5})(x^2+2)$, $(x^2-x-1)(x^2+2)$.

18. $\left(x+\frac{1}{\sqrt{2}}+\frac{1}{\sqrt{2}}i\right)\left(x+\frac{1}{\sqrt{2}}-\frac{1}{\sqrt{2}}i\right)\left(x-\frac{1}{\sqrt{2}}+\frac{1}{\sqrt{2}}i\right)\left(x-\frac{1}{\sqrt{2}}-\frac{1}{\sqrt{2}}i\right)$,
$(x^2+\sqrt{2}\,x+1)(x^2-\sqrt{2}\,x+1)$, x^4+1.

19. $\frac{1}{2}$, $-\frac{2}{3}$. **20.** $\frac{2}{5}$, $-\frac{1}{3}$, $\pm\sqrt{2}$.

23. All irred. except (iii).

Exercise 7B (p. 128)

5. H.C.F. (x^2+2); L.C.M. $(x^2+2)(x-1)(x-3)(x+5)$.

6. H.C.F. $(2x^2+1)$; $2(x+1)^2(x+2)$; $2(x+1)(x+2)(x^2+1)$.

7. H.C.F. $(x+2)^3$; $(x+2)^3$; $(x+2)^4$.

8. H.C.F. $(x+1)$; $(x+1)^3$; $x(x+1)(x+2)$.

9. H.C.F. $(x+1)(x+2)(x+3)$; $(x+1)(x+2)(x+3)$;
$(x+1)(x+2)(x+3)(x+4)$.

Exercises 8A (p. 148)

1. (i) $(1+\frac{3}{2}\sqrt{3}, \frac{3}{2}-\sqrt{3})$; (ii) $(\frac{5}{2}\sqrt{3}-1, \frac{5}{2}+\sqrt{3})$; (iii) $(\frac{1}{2}, -\frac{1}{2}\sqrt{3})$; (iv) $(\frac{1}{2}\sqrt{3}, \frac{1}{2})$; (v) $(-\frac{3}{2}-\frac{7}{2}\sqrt{3}, \frac{3}{2}\sqrt{3}-\frac{7}{2})$.

5. $\frac{1}{2}(\mathbf{a}+\mathbf{b})$.

7. (i) Perpendicular bisector of AB; (ii) circle centre A radius λ; (iii) circle (Apollonius's circle) on CD as diameter, where C, D divide AB in the ratio 2:1 and −2:1.

12. (−1, 5), (−4, 10), (−4, −3), (7, −2), (−7, 2).

18. (0, 0), (1, 0), (0, 1), (1, 1).

19. (0, 0, 2), (2, 4, 3), (4, 0, 4).

Exercises 8B (p. 149)

2. (i) $\lambda\mathbf{a}+(3-\lambda)\mathbf{b}+(5-\lambda)\mathbf{c}$; (ii) not possible; (iii) $\lambda\mathbf{a}+(1-\lambda)\mathbf{b}+(1-\lambda)\mathbf{c}$; (iv) not possible.

Exercises 9A (p. 167)

1. (i) No; (ii) Yes.

2. Yes. **3.** No. **4.** No. **5.** No. **6.** Yes. **7.** No.

8. Yes.

9. Onto and 1-1. **10.** Onto. **11.** 1-1. Positive reals.

12. Onto. **13.** Neither. $A\theta$ is the set $1 \geqslant t \geqslant -1$.

14. Onto and 1-1. **15.** 1-1. $A\theta$ is set $1 > t > -1$.

16. Neither. Image is set of all fathers.

17. Neither. Image is set of all eldest children. **18.** Onto.

19. Onto and 1-1. **21.** 1907, 1932.

Exercises 10A (p. 191)

11. 10 only.

14. (i) $c-b-a$; (ii) $-a+c-b$; (iii) $-b-a+c$.

16. True. **17.** True. **18.** Not true.

19. True. **20.** Not true. **29.** True.

30. True. **31.** Not true. **32.** Not true.

33. Not true. **34.** True. **35.** Not true.

36. True. **37.** Not true.

47. $x = (c+d)(a+b)^{-1}$. **48.** $x = a^{-1}(c+d)b^{-1}$.

49. $x = -c+(a+b)^{-1}d$. **50.** $x = -b+a^{-1}gd^{-1}c^{-1}$.

Exercises 10B (p. 193)

1. (i) True; (ii) not true; (iii) true; (iv) true; (v) not true.

2. (i) Meaningless; (ii) true; (iii) true; (iv) true; (v) meaningless.

6. (i), (iii), (iv), (v), (vi), (viii) have a unity. (iv), (vi), (viii) have inverses. ((vii) has no unity and hence no inverses.)

Exercises 11A (p. 215)

1. Yes. **2.** No. **3.** Yes. **4.** No. **5.** Yes.

6. No. **7.** Yes. **8.** Yes. **9.** No. **10.** Yes.

11. No. **12.** No. **13.** Yes.

17. e has order 1; u, u^2 have order 3; w, wu, wu^2 have order 2.

18. e, wwu, wu^2 are their own inverses; u and u^2 form a pair of inverses.

Exercises 11B (p. 216)

1. ba, ab have order 6; ab^2 has order 3.

2. (i) $ab^2 = b^3a, ab^3 = b^2a, ab^4 = ba$;
(ii) all have order 2;
(iii) $(ab)^{-1} = ab = b^4a, (ba)^{-1} = ab^4 = ba, (ab^3)^{-1} = ab^3 = b^2a$.

3. (1) 6. e, a, a^2, b, ab, a^2b. (In fact $b^2 = e$ and $ab = ba$.)
(2) 10. $e, b, b^2, b^3, b^4, a, ab, ab^2, ab^3, ab^4$.

Exercises 12A (p. 253).

1. No. **2.** Yes. **3.** No. **4.** C_n. **5.** C_4.

6. No. **7.** C_5.

8. $R \times R \times R$ where is the group of reals. **9.** C_5.

10. No. **12.** $(1, -i, -1, i) \leftrightarrow (0, 1, 2, 3)$.

13. $x \leftrightarrow \log x$.

14. $e^{2k\pi i/n}$ with $0 < k < n$ and k prime to n.

15. (i) All non-zero elements are generators.
(ii) 1, 3, 5, 7 are generators; 4 has order 2; 2, 6 have order 4.
(iii) 1, 5, 7, 11 are generators; 6 has order 2; 4, 8 have order 3; 3, 9 have order 4; 2, 10 have order 6.

17. (i) C_8; (ii) C_{10}. **18.** 6, 7, 11. **24.** ba^2, ba^4, ba, ba^3.

26. a^3 has order 2; a^2, a^4 have order 3; a, a^5 have order 6; ba^r has order 2 for $r = 0, 1, \ldots, 5$.
$(a, a^5), (a^2, a^4)$ are pairs of inverses; the rest are their own inverses.

28. (i) 2, even; (ii) 4, odd; (iii) 2, odd.

29. (i) 2, even; (ii) 4, odd; (iii) 5, even.

30. (i) $\begin{pmatrix} 1 & 2 & 3 & 4 \\ 3 & 2 & 4 & 1 \end{pmatrix}$; (ii) $\begin{pmatrix} 1 & 2 & 3 & 4 \\ 2 & 4 & 3 & 1 \end{pmatrix}$; (iii) $\begin{pmatrix} 1 & 2 & 3 & 4 \\ 2 & 4 & 1 & 3 \end{pmatrix}$;
(iv) $\begin{pmatrix} 1 & 2 & 3 & 4 \\ 4 & 2 & 1 & 3 \end{pmatrix}$; (v) $\begin{pmatrix} 1 & 2 & 3 & 4 \\ 2 & 1 & 3 & 4 \end{pmatrix}$.

32. (i) $\begin{pmatrix} 1 & 2 & 3 & 4 & 5 \\ 1 & 4 & 2 & 5 & 3 \end{pmatrix}$; (ii) $\begin{pmatrix} 1 & 2 & 3 & 4 & 5 \\ 2 & 4 & 3 & 5 & 1 \end{pmatrix}$;
(iii) $\begin{pmatrix} 1 & 2 & 3 & 4 & 5 \\ 1 & 5 & 3 & 4 & 2 \end{pmatrix}$, (iv) $\begin{pmatrix} 1 & 2 & 3 & 4 & 5 \\ 2 & 3 & 5 & 4 & 1 \end{pmatrix}$, (v) $\begin{pmatrix} 1 & 2 & 3 & 4 & 5 \\ 3 & 4 & 1 & 5 & 2 \end{pmatrix}$.

33. (i) Odd; (ii) even; (iii) odd; (iv) odd; (v) odd.

35. (i) 2, even; (ii) 3, odd; (iii) 5, odd; (iv) 9, odd; (v) 7, odd.

37. (i) Vierergruppe; (ii) Vierergruppe; (iii) D_{10}.

38. Vierergruppe $(C_2 \times C_2)$.

39. $C_2 \times C_2 \times C_2$.

40. C_2. **41.** D_8. **42.** S_3. **43.** $S_3 \times C_2$. **44.** D_8.

48. Generators a, b; $a^4 = b^2 = e, ab = ba$.

49. Generators a, b, c; $a^2 = b^2 = c^2 = e, ab = ba, ac = ca, bc = cb$.

Exercises 12B (p. 257)

1. 3, 5, 6, 7, 10, 11, 12, 14 have order 16; 2, 8, 9, 15 have order 8; 4, 13 have order 4; 16 has order 2; 1 has order 1.

(3, 6), (5, 7), (10, 12), (11, 14), (2, 9), (8, 15), (4, 13) are pairs of inverses.

Exercises 13A (p. 284)

1. Yes. **2.** No. **3.** No. **4.** No.

5. Yes. **6.** No. **7.** Yes. **8.** No.

15. $\{e\}$, $n \geqslant 3$, n odd; $\{e, a^{n/2}\}$, $n \geqslant 3$, n even; D_{2n}, $n = 1$ or 2.

16. $\{e\}$, $n \geqslant 3$; S_n, $n = 1$ or 2.

17. The subgroup of reals.

18. $\{e\}$. **32.** No.

INDEX